T0226977

CCTV for Security Professionals

CCTV for Security Professionals

Alan R. Matchett

BUTTERWORTH
HEINEMANN

An Imprint of Elsevier
www.bh.com

Amsterdam • Boston • London • New York • Oxford • Paris • San Diego
San Francisco • Singapore • Sydney • Tokyo

Butterworth–Heinemann is an imprint of Elsevier

Copyright © 2003, Elsevier All rights reserved.

No part of this publication may be reproduced, stored in a retrieval system, or transmitted in any form or by any means, electronic, mechanical, photocopying, recording, or otherwise, without the prior written permission of the publisher.

Permissions may be sought directly from Elsevier's Science and Technology Rights Department in Oxford, UK. Phone: (44) 1865 843830, Fax: (44) 1865 853333, e-mail: permissions@elsevier.co.uk. You may also complete your request on-line via the Elsevier homepage: http://www.elsevier.com by selecting "Customer Support" and then "Obtaining Permissions".

All trademarks found herein are property of their respective owners.

∞ Recognizing the importance of preserving what has been written, Elsevier prints its books on acid-free paper whenever possible.

Library of Congress Cataloging-in-Publication Data

Matchett, Alan R.
    CCTV for security professionals / Alan R. Matchett.
        p. cm.
    Includes index.
    ISBN-13: 978-0-7506-7303-7   ISBN-0-7506-7303-6 (alk. paper)
1. Closed-circuit television. 2. Television in security systems. I. Title.

TK6680 .M28 2003
621.389′28–dc21

                                                                2002038613

ISBN-13: 978-0-7506-7303-7
ISBN-10: 0-7506-7303-6

British Library Cataloguing-in-Publication Data
A catalogue record for this book is available from the British Library.

The publisher offers special discounts on bulk orders of this book.
For information, please contact:

Manager of Special Sales
Elsevier
200 Wheeler Road
Burlington, MA 01803
Tel: 781-313-4700
Fax: 781-313-4882

For information on all Butterworth–Heinemann publications available, contact our World Wide Web home page at: http://www.bh.com.

Printed and bound in the United Kingdom

Transferred to Digital Print 2011

In fond memory of William D. Norman, QSM, my mentor and best friend.

Contents

Preface ix

Introduction xiii

1 Purpose of a Camera System 1
2 Determining Necessary Requirements 25
3 Key Elements of the System 35
4 Choosing the Cameras 57
5 Control Equipment 89
6 Recording and Video Storage 111
7 Enhancing Recording Capabilities 137
8 Covert and Overt Cameras 157
9 Connectivity 167
10 Outdoor Considerations 187
11 Tying It All Together 209

Appendix Common Terms Used in CCTV 231

Index 267

Preface

As a Security Design and Analysis Consultant with a technical background, I have read some wonderful and extremely informative books on closed circuit television (CCTV) systems. All of these books have proved quite helpful in my early years in learning how the cameras, lenses, and other components operate. One thing most of these books have in common, though, is that they are all very technical in nature.

To me, one noticeable gap in available security books was this: there was no book that explained how to determine what is important in a facility camera system for the security manager. There has always been somewhat of a communication gap between the installation organizations and the system's end user. This book was designed in the hopes of assisting both sides in the quest for improved security and customer satisfaction.

This book will take a conceptual look at CCTV systems, not from a technical viewpoint but from a functionality viewpoint. Looking at the design and analysis of a system as well as the functionality of a system's components, this book should be helpful to a wide range of security professionals. Security managers and practitioners should gain a better understanding of the fundamentals of a system and how to choose what features are necessary. Installers and technicians should gain a better understanding of how cameras and control equipment should be set up and why, from the customer's point of view. Security sales personnel should gain a better understanding of what is important to their customers and how to make sure that they are completely satisfied with the system.

If the salesperson uses the concepts covered in this book, he or she should be able accurately to determine exactly what the customer wants and expects from the camera system. It should help the salesperson to determine the best way to explain the system to the user and ensure that he or she is getting the expected functionality. Many times when the installing company has completed their work, the user will ask about functions that are not there. Usually the user is under the impression that they are getting features that are not always included.

When end users, security managers, installers, and sales people are brought up to the same level of system knowledge, all of the groups will have a much easier time understanding each other and the system with which they are working. The system can be configured and programmed in a manner that satisfies all parties, yielding a much better chance of customer satisfaction, much less chance of improper equipment setup, and much less chance of requiring additional trips to the site by installers and service personnel.

To understand the differences between the two groups we need to look at the training and duties of each side. Installers, sales people, system designers, and all those associated with the sales and installation of security equipment, video or otherwise, learn their trade from a technical, product-oriented perspective. To most along this career path, a CCTV system is a group of components connected together for proper operation. Each component has its own specifications and assigned tasks that it must perform to make the system operate properly. When talking to someone who has spent years with manufacturing and installation organizations, it can be difficult to explain the concept of a system. As each portion of the concept is revealed, a technically-oriented person will associate a function with a piece of equipment instead of analyzing the entire functionality that it is hoped to achieve. While it is a fundamental requirement of a good designer or technician to understand the full capabilities of the equipment, this problem reminds me of the old saying, "Can't see the forest for the trees."

Security managers and those who are end users, however, learn and operate from an entirely different perspective. While it is important that most have some level of technical comprehension, most of their job functions and training have been goal or concept oriented. While the installer's training is usually mostly hands-on electronically-oriented schooling, security managers tend to be educated in more of an assignment-study-test format. Traditionally this means that they must understand "the big picture" and are left to their own means of achieving the desired results. Differences become more obvious when a security manager and an installer describe the same system. A security manager's description may be: "This is a color camera system with coverage of the lobby, all exits, the parking lots, hallways, and storage area. There are two operator workstations where any of the cameras can be viewed. All of the video is recorded and two tapes must be changed every day." An installer's description may be: "This system consists of 21 indoor color cameras and nine outdoor color cameras. Indoor cameras are in light duty, wall mount enclosures with auto iris lenses and cover the lobby, hallways, and exits. Outdoor cameras are in environmental housings with heaters and blowers. All outdoor cameras are roof mounts and four are pan/tilt and zoom units. All cameras are connected through two multiplexers and recorders, which are bussed together and can be operated from two separate locations." As you can see, the installer's description is much more detailed, and centers on the components, while the security manager just describes the details of what is covered. The security manager would probably view the installer's description as more detailed than necessary, and the installer would probably view the manager's description as too vague.

While the technical workings of cameras, lenses, and cabling are very important to someone in the installation and service industry, the average security manager could not care less whether a 1/2-inch or a 1/3-inch format camera is used. What is important to that security manager is adequate coverage and good picture quality. There are quite a few resources available explaining the differences between cameras and lenses, but there has not been one that tells the security manager how to design and set up the system to obtain the view and information that are necessary. Often I am called in to an organization that is having problems with their camera system or where the system is not capable of what they thought it would be. What I have found in most cases is that the security manager had a basic idea of what was wanted from the system, but didn't know what guidelines to establish for the *installation company*. In most cases, the installation company was relied upon as the expert, but sales people and technicians are also not always familiar with the requirements, concerns, and liabilities of a security manager's position or the inner workings of that particular organization. It is up to the security manager to explain exactly what is needed in a system, what is expected, and what can be accomplished. Without this information, the sales and design staff of the installing company must make assumptions that may or may not be correct, or they may misunderstand what the security manager is asking for.

When designing a camera system for a client, I always start by finding out how they imagine the system working for them. We discuss their coverage and recording requirements, how they plan to manage the system, and what different levels of operation they may require. By taking the analysis from operation concepts down to specific details, a complete camera system can be designed that is sure to meet the customer's needs. In Chapter 2 we will discuss in more detail how to start with the concept and determine their system requirements.

After years of honing my own skills, studying design and installation techniques, and experimenting with different designs and system formats, I felt that it was time to fill the gap in security texts by compiling what I've learned and making it available to other security professionals. I decided to write this book in the hopes of training security managers to determine what is needed for the corporate CCTV system, and how to convey that information in the best way to the installation and sales people. It was also my goal to help train the sales and installation staff in the best way to help guarantee customer satisfaction. While many security managers may wish to let these companies decide their own fate, a better understanding of these systems from both groups will help to make sure that they are getting what they want.

By understanding the needs of their facility, managers can more effectively convey the necessary information to the installation company. This gives the installation company a much better chance of meeting the customer's expectations and of having a better long-term client relationship.

It is impossible to cover every type of facility and every system component, but this book makes every attempt to cover the essential elements of a system for a variety of businesses and applications. Each business, however, will have its own unique situations and needs, which may require a combination of principles from several facility types. When in doubt, the security manager should discuss the design with a consultant or system expert, hopefully with a better understanding of his or her needs and what questions to ask.

When reading this book, the reader may notice that some principles of a system seem to be repeated throughout several chapters. Picture resolution, for example, is covered in Chapter 4, "Choosing a Camera," as well as in Chapter 5, "Control Equipment." There are two reasons that some information is repeated throughout the book. First, factors like picture resolution can be affected by several pieces of equipment and each interrelates with the other. Just choosing high-resolution cameras will not make a high-resolution CCTV system. The second reason for some redundancy is so that this book can be best utilized as a reference tool. The reader should not have to read every chapter just to discover everything that affects picture quality if his or her biggest concern is with a time lapse recorder. A reader should be able to go to any chapter in this book and gain some insight into a camera system.

For those who decide to get this book and read it cover to cover, first of all, thank you! Secondly, if you find yourself feeling like you have already read a section, keep going, there will be new material covered along with the repeated information.

Introduction

This book is designed to be a valuable tool for security professionals. For the security manager, it should provide valuable information needed to properly plan and design a camera system to meet the needs of the organization. It should also give the security manager the knowledge to discuss the system, features desired, and areas of concern of the security sales and service organizations who may be tasked to sell, install, and/or service the system. If security managers are not in the position to design the system, they should gain a much better understanding of the components and functions of a system. This can prove to be invaluable when evaluating an existing system or evaluating a system proposed by a vendor.

Installers and service personnel should also benefit from this book. It should provide them with a better understanding of what their customers want, how to set up the system properly, and for newer technicians, what the functions and possibilities are with the various components of the system.

For security equipment sales personnel, this book should provide a better knowledge of the customer's needs, as well as an understanding of their point of view and reference. Often the sales person is provided with very little detail about what the user would like to achieve with the system and must try to put together a system that they feel would be beneficial. This book should help them determine exactly what questions they should be asking the customer and what the responses mean. It should show them what choices to give in order to help ensure that the customer is satisfied with the system and getting exactly what he or she expects.

In general, this book looks at camera systems from a different angle than most other books on CCTV. While most books look at the technical aspects of CCTV components, this book looks at camera systems from a user and functionality standpoint. Hopefully, it is written in a way that is both easy to understand and informative. While some sections do look at the technical aspects of equipment, it is done to provide a better understanding of the functionality and capabilities of that equipment.

While the chapters are in a sequence that guides the reader from basic principles to a completed system, they are also designed to be a stand-alone source on their respective subjects. It is hoped that in this way readers can use this book as an easy reference for years after they have finished their first reading.

Chapter Information

Chapter 1: Purpose of a Camera System
Chapter 1 starts with a look at some basic terms and functions of a camera system. Important attributes of a system are explained by looking at specific types of facilities. Comparing retail, hotel, educational, industrial, and office facilities, it becomes clear that one size does not fit all when laying out a camera system.

Chapter 2: Determining Necessary Requirements
This chapter looks at some of the planning that is necessary to decide exactly what the user hopes to gain with the system. By evaluating the facility and potential threats, the designer can determine what features are most important. Future expansion needs and the available budget can then be reviewed to help choose the control equipment and camera layout necessary to complete the system.

This chapter emphasizes the importance of planning and evaluation to the future usefulness of the system. A system designed by choosing equipment, then trying to get the most out of it, may be sufficient for some security managers, but a system that is designed based upon the true needs of the facility will address the user's specific system requirements.

Chapter 3: Key Elements of the System
Chapter 3 looks at the different functions of a camera system. This is an explanation of the most fundamental uses of a CCTV system, which are often taken for granted. It is not often that a system user looks at the simple act of viewing the camera activity or recording and playing back the video.

There may not necessarily be a right and wrong way to look at camera views on a monitor, but there are efficient and inefficient methods. Recording can be much more crucial to understand because improper recording can mean completely missing an important event.

This chapter ends with a look at the importance of handling recorded images in the case of a major event. Establishing a proper chain of evidence can be the deciding factor for whether or not a video clip or tape is admissible in a court of law.

Chapter 4: Choosing the Cameras

Using the preliminary research from the preceding chapters, Chapter 4 looks at the information needed to choose the correct cameras.

Deciding whether the system will be color or monochrome is one of the most fundamental decisions that must be made before most of the components can be selected. Chapter 4 looks at the advantages and disadvantages of both.

Understanding camera specifications is very important when choosing a camera. A degree in engineering should not be required to understand which camera is best for a particular application. Hopefully this section will help to clarify the specification terms without being too technical.

Lens selection goes hand in hand with camera selection and can be very confusing. This chapter breaks it down to a much simpler process that is fairly accurate.

Other camera system components are often necessary at some locations, so this chapter also looks into selecting zoom lenses and pan and tilt units.

Chapter 5: Control Equipment

From a simple single camera system to complex systems with hundreds of cameras, control equipment makes the camera signals useable for the system user. Chapter 5 covers the various types of control equipment, how they are used, and how they affect other components.

Control equipment includes many components to view, manipulate, control, and improve the video received from the eyes of the system, the cameras. In this chapter the reader will learn how to pick the correct items for viewing multiple cameras from single or multiple locations, record more than one camera on a single recorder, move cameras to different viewing areas, and enhance the image and recording quality to make the video more useful.

Chapter 6: Recording and Video Storage

Chapter 6 compares the different types of video recorders available, covering probably one of the hottest topics in the security industry. This chapter presents an in-depth description and comparison of the various types of analog or traditional video recorders explaining the fundamentals of digital recording. The digital recording section looks at the types of recording available and the various ways that digital can improve system operation. Disadvantages of digital record-ing will also be covered as well as the methods of viewing the video over a net-work, via modem, or even over the Internet.

Chapter 7: Enhancing Recording Capabilities

Whether the reader chooses traditional or digital recording methods, it is impor-tant to get as much useful recorded video as possible. Chapter 7 discusses some techniques, programming features, and equipment that ensures that the recorded images are useful and not just time consuming. The reader will learn about effective use of alarm triggers, video motion detection, and output relays to improve video recording. This chapter also looks at effective ways to interface the camera system with alarm systems and access control systems to improve the overall security of the facility.

Chapter 8: Covert and Overt Cameras

Though relatively brief, Chapter 8 discusses applications and problems with using covert or hidden cameras. Improper use of hidden cameras can be more than just poor judgement—it can be criminal. Therefore, this subject deserves a chapter of its own.

Chapter 9: Connectivity

Chapter 9 looks at the various methods used to get the video signal from the camera to the control equipment effectively. In addition to looking at the various cabling options, this chapter also looks at the types of connectors commonly used and some wireless video applications. This section should provide the reader with a better understanding of coaxial cable, fiber optic cable, twisted-pair cabling systems, and wireless video transmission. It should also help to explain the differences between certain connector types and advantages or dis-advantages of the various interconnection methods.

Chapter 10: Outdoor Considerations

This chapter takes a closer look at variables that affect camera operation and video quality when cameras are installed outside. Climate, lighting, and camera accessibility can affect not only long-term operation of a system, but also future maintenance costs and serviceability. The reader will learn what simple, com-mon mistakes to avoid for an improved camera system.

Chapter 11: Tying It All Together

Chapter 11 uses the information learned in the previous chapters to create a system for an imaginary facility. This chapter will give the reader a practical look at doing an actual system design. Using fundamentals that work in most design applications, the reader will see the design take shape from two distinct vantage points, looking first at the functional requirements and then at budgetary caps. With the budgetary scenario the reader will look at the most effective way to design a system, allowing for future budgetary considerations.

CCTV for Security Professionals

Purpose of a Camera System

BASIC TERMS AND CONCEPTS

To best understand the needs and possibilities of a modern camera system, some terms and concepts must first be understood. The phrase *closed-circuit television video system* is abbreviated as CCTV. Broadcast video such as traditional television is a much more complex system in which the video images and audio tracks are converted to signals, which are transmitted and can be received by anyone within range who has the proper equipment tuned to the proper frequency. Because the video is available to anyone who is tuned to the correct frequency, broadcast television is essentially an open-circuit system.

With CCTV systems, however, signals from a video source, such as a camera, are transmitted by a direct connection to the receiving equipment, such as a monitor. This connection is usually made with coaxial cable but can also be made with fiber-optic cable or a single twisted-pair cable if the correct conversion equipment is used. This connection makes a completed closed circuit, which cannot easily be viewed from outside of the system. In theory, the only way to view the video images from the camera is with a piece of equipment that is part of the closed circuit. Modern camera systems are not always a true closed-circuit system, although they are still usually referred to as such.

Many new devices, such as wireless transmitters, Web cameras, and camera servers, are being added to camera systems now. These devices allow people with the proper knowledge and/or equipment to receive the video images from the camera without being directly connected to it. Although theoretically this is not a closed-circuit system, it has drastically expanded the applications and usefulness of video surveillance systems. Day care centers, for example, can now provide a means for their clients to check in on their children and the facility at any time remotely via the Internet.

The role of the camera system in a security program has evolved substantially in the last decade, as has the equipment. In the past, cameras were thought of as an extreme measure, primarily for banks and large office buildings. Camera systems were mainly used in facilities with a 24-hour guard service. Most systems consisted of a few cameras, monitors for the guards to view, and possibly a switcher to view multiple cameras on a single monitor. These systems were not considered practical for most facilities because they were fairly costly and only beneficial for live viewing.

Advanced technology brought the video recorder and eventually the time-lapse recorder, which made it possible to store images for later review of any incidents that may have occurred. The recording devices expanded the possible uses for camera systems, but recorder costs still limited their use. In the 1980s, many new products and shrinking equipment costs greatly expanded the applications and use of camera systems. Many households now have video recorders, and the lower price of manufacturing has spread to the time-lapse recorders used in the security industry.

Video multiplexing and the ability to record multiple cameras to a single tape further expanded the growth of system installations, probably more than any other CCTV item made. Before the multiplexer, a switcher could be used for viewing multiple cameras, but the recordings were still only of the image that was being displayed. This meant that with eight cameras on a switcher, and each camera displayed for just one second, each camera would be recorded and displayed only once every eight seconds. Seven out of eight seconds from every camera location were missed, so there was more likelihood to miss an event than catch one. Even when an event could be captured, the amount of usable video meant that most of the evidence was missed and what was there could be open to more than one interpretation of what was actually happening.

By adding a multiplexer to the system, much more video from each camera could be recorded onto a tape than with a switcher. The elapsed time between each recorded image would be much quicker than with a switcher, providing more useful recorded video. Another advantage to adding the multiplexer was that multiple cameras could be viewed on the monitor at the same time. This meant fewer monitors, an easier time for guards to view multiple cameras, fewer recorders, and less time to review video footage.

As more and more equipment was invented to enhance the CCTV system, the system's role gradually changed. Today video is used in virtually every type of facility, and in many cases has become a necessity. With all of these changes and advances in the video industry, it is still important to keep the primary purpose of the camera system in perspective. A camera system cannot protect people from crime. Cameras cannot protect property from theft or vandalism. The camera is merely the silent observer, watching what it has been set up to watch. The recording device is there to gather and store information, primarily for

Figure 1.1 The camera view of an entrance door can be closely monitored by a guard.

future use as evidence if needed—evidence for a court of law if necessary or company evidence for confronting dishonest employees or patrons should that need arise. Camera systems are merely a tool that, when set up properly, can detect activities. These activities must still be interpreted by a human being, who can then take whatever actions may be necessary as a means of response.

For example, a camera may be set up to view a door at the rear of a facility (see Figure 1.1). A guard, who is assigned to monitor the activity on the cameras, may notice someone attempting to gain entry into the facility through the rear door. The guard must then interpret the images that he or she is seeing and decide what action must be taken in response. If the person is attempting to break into the facility, the guard may wish to call on the guard force to apprehend the person. If, on the other hand, the person attempting to gain entrance is just an employee with the proper authority to use this entrance, the guard may choose to take no action at all. The video images in this case were a tool that allowed the guard to analyze what was occurring at the rear door without actually being there. The camera system itself has no way of interpreting this data to take the appropriate action. Alarm triggers and relay outputs could be used to initiate a response if the door was actually opened, but the actions must still be interpreted by a human being.

If the camera system is set up to record the video images, it can provide an additional source of evidence if an incident occurs. The camera system, however, has no means of determining what is actual evidence and what is not. It will merely record what it is told to when it is told to. A human being will then be able to review and analyze the information that has been recorded and take the appropriate action.

One point for heated debate within the security industry is the use of camera systems as a crime deterrent. Visible cameras, mounted in the most obvious places, were considered to be a good deterrent for many crimes, such as armed robbery, burglary, and shoplifting. Although many security professionals may still argue that cameras are a great deterrent, this is not—and should not be—the primary purpose of the system. Many cities have begun installing cameras citywide based on the assumption that cameras will discourage people from performing criminal acts. The cameras may discourage some activities from happening in the immediate viewing area of the cameras, but there has not been enough evidence gathered to show that the cameras deter the crime from happening altogether (see Figure 1.2).

While it is not the point of this book to editorialize on the validity of this theory, it should be mentioned as another current use of video systems. Necessary design techniques should be mentioned for systems utilized for this purpose, whether or not each reader believes that this is a legitimate application for video surveillance.

Camera systems may not deter crime, but they do seem to at least defer crime to other areas. If the crime still occurs but is covered by video surveillance, it is hoped that the system has been set up properly to capture the event for future use as evidence. Home intrusion detection systems are similar. They may not make a criminal decide to stop breaking and entering but instead search for an easier target. With video systems the same is true, which gives the security manager more of an ability to control his or her environment.

Insurance companies and the judicial system have also established camera systems as a necessity to limit or decrease potential liability. From a corporate point of view, a camera system in conjunction with other security measures, such as lighting and alarm systems, can show that a company has taken adequate and reasonable measures to protect employees, customers, and clients.

One important consideration to remember with camera systems in a business environment is the employees' attitude toward the system. Immediately

Figure 1.2   One area for debate among security professionals is whether visible cameras deter crime or simply defer it to other areas not covered by cameras. If either is correct, it can be very beneficial for a security manager responsible for the safety and security of a building.

after a camera system is installed, employees often have mixed feelings about having cameras everywhere. There are sometimes feelings of relief that cameras are being installed, but usually there are concerns that the employer does not trust the employees or will use the camera system to make sure no one is goofing off. Therefore, it is recommended that the purpose of the system for that facility is briefly explained to everyone. For example, a memo could be sent to all employees that states: "A camera system has recently been installed as a means of enhancing the security of the facility. This new system will be used to help provide better security for our employees and our equipment. If you have any questions or concerns about the new camera system, please ask the security manager." Make sure you have any such memo approved by the company legal adviser and be honest with employees. That is the best way to alleviate their fears.

Also keep in mind that after the system has been in use for a while, most employees tend to forget about it and continue as if it were not there. After a few weeks, the cameras become part of their daily environment and no longer something to be concerned about.

These may be the general reasons to incorporate a camera system into a security program, but the specific purposes of the system can vary according to the type of facility using the system. Factors such as size and type of system used, camera locations, recording capabilities, and monitoring requirements all vary greatly from one facility type to the next. Video applications for each environment could easily fill a separate book, but some key factors of each are covered here.

CAMERAS FOR RETAIL SECURITY

Camera systems have become a large part of the retail environment. System uses and requirements can vary as greatly as the types of retail stores. Stores that sell large items such as furniture and appliances, for example, would not be nearly as concerned about shoplifting as clothing stores and music stores. Although shoplifting may occur occasionally, it is unlikely that someone will try to walk out of a store carrying a couch. Clothing, on the other hand, would be easy to conceal and much more tempting to take (see Figure 1.3).

In a retail environment, the basic purposes of a camera system are the protection of products and the protection of people. The protection of products is somewhat more extensive in the retail environment than in other facility types. All businesses are established to sell some type of product, even if that product is a service, but the retail industry is probably the most vulnerable to product theft. In a retail store, more people have access to the products everyday than in most other environments. Patrons are there solely because they have an interest in the products, and employees are always around the items and familiar with their handling.

With such a high volume of people and products, it is nearly impossible to watch all of them all of the time, which usually provides ample opportunity for theft. Such a high volume of products and potential customers means a higher number of employees than many other business types. These employees, often driven by greed, desire, anger, or many other reasons, are the most likely to take products illegally. This statement may seem unfair, but loss prevention statistics consistently show that most loss occurs internally.

Figure 1.3 Retail checkout lanes often require many cameras in a small area.

In addition, with large numbers of employees and patrons comes an increased risk of lawsuits and claims against the retailer (see Figure 1.4). From discrimination to personal injury, such as from a slip and fall, the camera can assist in proving or disproving any allegations. When utilized properly, the camera system can be a beneficial tool for creating a safer workplace and protecting people. Watching at all times, the camera system can be used to help protect patrons from store employees and other patrons, employees from hostile customers and coworkers, and the retailer from all of the above.

Other areas of concern are items that never actually leave the store. In this case, a person enters the store and grabs several items of substantial value. Rather than attempting to walk out of the store with them, he or she takes them to the counter as a return. This person may attempt to get cash back for these items or store credit to be used then or later. Unless people are observed coming into the store empty handed, picking up the items while in the store, and taking them to the returns counter, it can be difficult to prove that the item is not a legitimate return. A store policy that receipts are not required for returns can make it much easier for someone to cheat the retailer. For stores that have a no-receipts-required policy, the camera system becomes much more essential in detecting fraudulent returns.

The number of scams that are attempted in retail establishments can be practically endless. Price tag switching, concealment of one product inside another, attempting to confuse a cashier, altering checks, using stolen credit cards, and shoplifting are just a few. Two fairly recent means of theft should be noted. Smash-and-grab robberies are becoming more commonplace worldwide. In this scenario, the theft usually occurs when the store is closed. A car, usually stolen, is rammed through the front of the store. The thieves run through the newly made entrance and grab as much as they can, often targeting specific items that can most easily be sold later. They then flee to a second waiting vehicle and are gone in minutes, well before any police response can be made. The

Figure 1.4 Warehouse areas of a retail facility have multiple reasons for complete camera coverage. Employee theft and safety concerns, for example, may make it necessary to monitor all exits and traffic areas where forklifts are operated.

camera system becomes important in this type of situation because obtaining eyewitnesses and actually catching the persons in the act are highly unlikely options.

Another variation of the smash-and-grab is the daytime grab-and-run. In this scenario, a large group of people enter the store simultaneously. By numbers and the element of surprise, store clerks usually have no idea how to react. Each person from the group grabs as much as he or she can and runs out of the store. No attempt is made to sneak items past security measures; this is just grabbing what he or she can carry and running before anyone can react. Because several people are usually involved, it can be difficult for eyewitnesses to give accurate descriptions. The sheer speed of the event and number of people can leave store clerks and customers confused about who did what. Another retail environment, gas stations and convenience stores, consistently tops the charts for armed robberies. Therefore, camera systems have become much more commonplace in this type of environment. Additional concerns in this environment include shoplifting and drive-offs, where people fill their vehicle with gas and leave without paying for it.

Retail Design Considerations

When designing a camera system for a retail environment, there are several important design considerations to keep in mind. It should first be determined whether color or monochrome cameras will be used. Although both may be effective, there are some distinct advantages to the color camera system. Monochrome cameras often provide somewhat higher resolution for each picture, but the color system can usually provide better identifying characteristics. Color variations of products, vehicles, and hair and skin tones help make future identification much more reliable in a court of law. A wide variety of products may appear in the camera viewing area at any one time, many of which are similar in size and shape. Although the monochrome camera may be useful in identifying activity, such as movement from one area to another, the color camera can be much more precise in distinguishing people or objects when set up properly.

Another important consideration in a retail establishment is camera location. Cameras should be installed to cover the entrances and exits, cash registers, and checkout areas. A detailed site analysis and traffic analysis should be performed to determine what areas of the store will need the most coverage and where a person could most easily be identified in the act of committing a crime. When a loss prevention staff member will be monitoring the cameras at all times, there are some areas in which cameras with pan, tilt, and zoom capabilities can

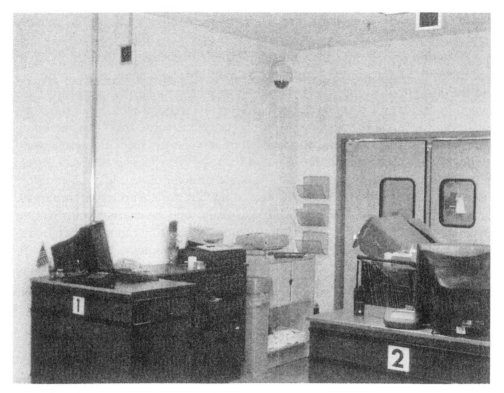

Figure 1.5 Proper camera location is essential to monitor a retail facility. This pan/tilt zoom dome camera allows the security staff to change the view as needed in the returns area of a store.

provide the most benefit. These would be areas such as cash registers, where it may be most important to zoom in on a person to see specific actions or movements (see Figure 1.5).

The pan/tilt zoom cameras should also be used in locations where it may be important to track a person's movement through the store until he or she is picked up by another camera. In the retail environment, when using video surveillance as evidence, it is often important that the suspect remains in full view of the cameras or a loss prevention officer at all times. Cameras should also be installed to cover the most vulnerable products. Although a dining room table may have a much higher dollar value, it is not as vulnerable to theft as a compact disk or computer software.

Because of the many types of loss that can occur, the recording capabilities in a retail environment can be one of the most important factors. If the frame rate of recorded images is too slow, important evidence can easily be missed. On the other hand, the faster the recorded frame rate is, the more frequently the tape or other recording media must be replaced. The frame rate refers to the number of images per second that are recorded onto a tape. With a traditional videocassette recorder (VCR), for example, the frame rate is typically 30 frames per second, whereas with a standard time-lapse recorder, the frame rate is typically 5 frames per second.

Because loss in most retail environments occurs during normal business hours, this is when the best recording capabilities would be required. Viewing each camera once every two seconds or more would probably be too slow to accurately capture an incident. By selecting a faster recording speed, such as 8 to 12 hours per tape, however, the number of images per camera per second can be increased to a much more desirable level. The benefits gained by using a faster recording rate should far outweigh the inconvenience of changing tapes more frequently and using more tapes.

For after-hours recording, a lower recording range of 24 or even 72 hours per tape can effectively be used. After hours, recording should still be done, but just being there when the store is closed is potentially a crime, so recording every movement may not be necessary. After hours, the video would only need to establish presence, objects being carried or stolen, and identity of the subject. During store hours, the video may also need to show the concealment of an item or hand activity at a checkout register, which can happen in seconds.

One efficient way of recording both times effectively with as little inconvenience as possible is to use two separate recorders. One recorder would be used to record the business hours activity and could be set up to use one or two tapes per business day, depending on the recording detail needed. The second recorder would be used to record activity only when the store is closed. Because a slower record time would be needed, this machine could be set so that the tape could be used to cover two or more nights without being changed. By using two separate machines and establishing specific recording and tape-changing procedures, confusion over recording speeds and tape-changing times can be reduced. Each tape can be used in its entirety as well, instead of using half a tape and stopping.

In many retail environments, especially with jewelry and banking, real-time recording should be available during emergency or alarm conditions. The ability to trigger this faster recording rate should be available to all tellers, clerks, checkout personnel, and security personnel. The system should continue to record in real time until the alarm is manually reset or the tape or storage space is full.

CAMERAS FOR HOTEL AND LODGING SECURITY

Camera systems in the hotel environment have become much more prevalent than in decades past. Hotels and other lodging facilities have a few distinct concerns that warrant the use of video surveillance as a means of enhancing security. Besides armed robbery of hotel clerks, there are a few specific and growing areas of concern in this atmosphere. Robbery of hotel rooms and attack of clients, for example, could easily be a cause for concern, leading to potential lawsuits. Personal injury cases are another liability that lodging establishments must face on a growing basis. One big area of concern for hotels is the safety of their customers and visitors in parking garages. Vast amounts of space, a large number of obstacles, and low lighting levels make these areas a prime location for auto theft and personal attacks on those coming from or going to their cars. These same conditions also make video surveillance much more difficult. In fact, some security consultants specialize in parking lot security. Because there are so many variables and areas of concern on this subject, it would be wise to work with one of these consultants if parking garage security is a must.

Hotel Design Considerations

For the hotel and lodging industry, safety and the protection of people are the two biggest purposes for adding a camera system. Color cameras are still recommended, especially in lobby and hallway areas, but monochrome systems can still be effective.

When determining camera location, some areas should not be overlooked. Building entrances with marble or potentially slippery floors are prime areas for people to slip and fall or to fake an accident. Sliding doors have also been known to cause injury and are also great for con artists looking for a quick settlement. Elevator lobbies and secluded public areas of the building, such as stairwells and emergency exits, should also be adequately covered. For parking garages, it is important to use cameras that are capable of providing adequate coverage with the lighting that is used. As in most situations, the camera system should not be the only source of security. Emergency call stations, intercoms, and fire pull stations should also be used in addition to video surveillance.

Recording speeds should be looked at closely when designing a system. Hotels are normally open 24 hours a day, 365 days a year, so recording should be done at all times. A recording rate of 24 hours per tape is usually sufficient. This rate can be adjusted for each facility and enhanced during nonpeak hours and in low-traffic areas with video motion detection and alarm triggers.

Because of the likelihood and high cost of personal injury lawsuits, I recommend maintaining a large archive of recorded videotapes. In the retail environment, the tapes can be reused on a weekly basis fairly adequately. For personal injury cases, though, it could be weeks or even months before a case is filed. In fraudulent cases specifically, the person filing suit may wait until he or she believes that a tape showing the incident no longer exists. It is far cheaper to purchase and store even six months of videotape than to have to settle a lawsuit because the alleged event cannot be disproved.

CAMERAS FOR THE OFFICE ENVIRONMENT

Although theft and property protection are concerns in an office environment, these issues are usually not as much of a priority as they are in a retail establishment. Retail facilities may have many people wandering throughout the facility, whereas an office environment is usually more structured and controlled. Most activity in this atmosphere is by employees, with the occasional visitor, guest, or contractor who usually must sign in or be escorted by an employee.

Most loss in an office environment is employee theft, traditional burglaries, and—with ever-increasing competition in many industries—industrial espionage. Of these loss possibilities, employee theft is the most common, probably more than most employers care to admit. For this reason, camera coverage is somewhat different from what is required in retail establishments.

In the not-too-distant past, theft would have been the biggest concern for a security or human resources manager, but the 1990s have brought a new leading concern to the forefront. Media coverage of violence in the workplace has increased tremendously in the last decade, making the protection of people a top priority with the camera system. Although it is not the intent of this book to focus on workplace violence incidents or statistics, it should be pointed out that this issue is more of a concern now than it was in the past for most companies.

Workplace violence can stem from a disgruntled employee or former employee, from one employee to another for various reasons, or from outside association of an employee, among other things. These incidents usually occur during normal working hours. Cameras in this situation can be used either to alert security personnel of the presence of a person who is a concern or to document an incident that may occur.

Office Design Considerations

When designing a camera system for an office environment, camera coverage areas are the top concern. Cameras should be located to cover all perimeter entrances and exits, ensuring that anyone coming or going is caught on tape. Hallways should be considered the next concern for coverage, as well as the lobby or reception area, if the facility has one. Because many incidents happen as people are coming from and going to work, parking areas and outside perimeter coverage is another concern. If a guard force is used, this coverage can also provide an early warning of potential problems. A person who is planning a violent incident will often stake out the facility several times to plan the attack or wait for the right opportunity, giving facility personnel ample documentation of trespassing or possibly stalking evidence.

Private offices within the facility are usually the areas of least concern. Although the office may be the property of the corporation, cameras installed in these areas could be deemed as an invasion of privacy. Laws and rulings regarding camera placement in offices vary from state to state, so if this is a desired measure, the security manager would be well advised to discuss the possible impact with company lawyers and law enforcement officials before anything is installed.

Color cameras are recommended for an office environment, although monochrome cameras will work sufficiently. Using color in an office environment is not as crucial as it is in other business types, such as gaming establishments. For most indoor camera applications, the lighting is sufficient to provide a good, clear color image with adequate resolution. It is important to find out in advance whether lighting will be on in the evenings at a level sufficient to support the camera requirements. Areas within the building that have low pedestrian traffic can also benefit greatly from the use of video motion detection or alarm triggers to reduce the amount of unnecessary video of an empty room or hallway.

For outdoor cameras covering the parking areas and building perimeter, monochrome cameras should be considered. Color cameras would be effective for daytime images, but nighttime could cause color cameras to show just a dark screen, depending on the light level. Color cameras are continuously improving, and the lighting required is getting much closer to that required for monochrome cameras. In most cases, however, the monochrome camera will provide superior nighttime images. An alternative is to use a day/night camera, which changes from a color camera to monochrome as light levels decrease. When the light increases to a sufficient level, the color camera takes over again. These cameras can be expensive, but the costs should be compared with adding additional lighting and using color and monochrome cameras to cover the same area.

Whereas the retail environment requires a higher picture-per-second rate of recording during hours of operations, office facilities will need excellent 24-hour recording, usually with alarm inputs and movement detection to enhance the frame rate. This topic is covered in more detail in Chapter 7, Enhancing Recording Capabilities. In general, while retail incidents usually involve some type of deception that can happen quickly, office incidents can usually be captured adequately with as little as one or two frames per second. With the proper combination of alarm activation and movement detection, 24 hours can usually be recorded on a single tape from multiple cameras during weekdays. An entire weekend from multiple cameras can also usually be put on a single tape, depending on the hours of operation for the establishment.

CAMERAS FOR EDUCATIONAL FACILITIES

School security is one of the fastest-growing areas in the security industry, and with good reason. Just as violence in the workplace has been on the rise, acts of violence in U.S. schools have increased dramatically. From petty theft to mass murders, what was once unthinkable is now occurring in our schools from high schools to elementary schools. Private guard forces and local police departments are being assigned to monitor schools more frequently. Metal detectors, camera systems, and school policies are being put into place to try to maintain control as more violent acts occur. Although the camera system cannot prevent a student from shooting another student, perhaps it can provide school administrators and guard forces with an early warning to stop such events.

If a camera system is going to be added to an educational facility, great care should be taken in advance to make sure this move is perceived as a benefit by faculty, students, school board members, and parents. It is recommended that the purpose of the system be openly discussed so that it is not perceived as martial law and/or an invasion of privacy. The purpose of the system should be not so much to monitor the activities of the students as to help protect the students and faculty. If the system is not accepted from the beginning by all parties as a benefit and a necessity, use of video surveillance could backfire and cause more disruption than benefit.

School Design Considerations

There are many areas of concern for camera coverage in educational facilities, depending on the budget size and purpose of the system specific to each school.

In general, the entrances and exits as well as the hallways should be adequately covered. In some instances it may be necessary to have camera coverage in each classroom. This system can be used to avoid allegations of teachers abusing students or students abusing teachers. In most cases, this undertaking is expensive and not feasible. If cameras are installed in each classroom, another important factor to consider is the location and accessibility of the control and monitoring equipment. If video images are transmitted from each classroom, the control and monitoring equipment should be extremely well protected and the location not publicly known. In a large-scale emergency, cameras in every classroom and control equipment accessible by emergency personnel could help greatly to locate and extract students and faculty. On the other hand, if camera views of all classrooms had been accessible by the Columbine gunmen, those people who survived by hiding in a locked room could have been discovered and possibly killed. Remember to consider all scenarios and possibilities that may occur when planning the system.

Recording requirements in a school environment are fairly straightforward compared with some other facility types. During school hours, to include extracurricular activities, a recording rate of 24 hours per tape or faster should be selected. In many cases, a recording rate of 12 hours per tape will effectively cover the entire time of the day that the building is occupied. Recording after hours may not be necessary at all, but recording can be set to occur in the event of an alarm trigger or video motion detection. This approach will keep unnecessary recording to a minimum but still capture any suspicious activity.

Camera choice can be important when designing a system for educational facilities. Vandalism and damage to cameras can be likely, especially in higher-grade levels such as a high school. Although vandalism may not be as common in elementary schools, it still does occur. It is much cheaper in the long run to start with vandal-proof cameras, mounts, and enclosures than it is to install and replace a less durable unit. Many schools have even gone so far as to install industrial cameras in ballistic housings, which are designed to survive direct weapons fire and explosive charges. Although this extreme may not be necessary for every school, the more durable the initial camera is, the less likely it will need to be replaced later.

All cabling and connectors should be completely concealed to prevent vandalism and tampering. As sad as it sounds, many designers for educational facilities use the same design approach as someone designing a system for a prison. Remember that if it can be broken, it will be broken, so start with durable equipment and a tamper-resistant installation.

CAMERAS FOR DAY CARE FACILITIES

Camera systems in day care facilities and "nanny cams" are a fairly new application for video surveillance. In this situation, the primary purposes of the system are to keep an eye on the children and the day care workers and for the parents' peace of mind. Camera systems in day cares are most likely not installed to watch for theft or vandalism. Several highly publicized cases of abuse by day care workers prompted many facilities to add video surveillance to watch the children at all times. Advancements in technology and the Internet have added the ability for parents to view the cameras remotely at any time to look in on their children. Many insurance companies and local licensing agencies have also begun making video surveillance or considering it for reduced insurance rates and licensing.

Day Care Design Considerations

Video cameras should be set to record at all times that any children are in the facility. Every area of the facility should be completely covered, with no blind spots into which a child could be taken. This does *not* include bathrooms, as with most facilities; however, any entrances into bathrooms should be adequately covered to see who takes a child in, when, and for how long. Covering the entire facility can mean a large number of cameras, sometimes two or three cameras per room. Adequate multiplexing and recording equipment should be installed to make sure that each camera view is recorded at least once per second.

Because many facilities can house numerous children and day care workers at any given time, color cameras should be used to make identification of any individual easier. Many facilities have begun to have the children signed in by their parents. Each child can then be assigned a numbered or lettered vest that he or she wears for the day to make it easier to identify each child on a monitor. For parents who can check in on the facility, it is also easier to locate their child and avoid any confusion. With the proper equipment setup, parents would be able to check on their children at any time via an Internet connection or by dialing directly into a camera server via modem from their computer.

CAMERAS IN A MANUFACTURING OR INDUSTRIAL ENVIRONMENT

Camera equipment that may be perfect for retail and office environments may not perform effectively in manufacturing facilities. For the manufacturing of technology products and some other things, there may be a laboratory-like environment with controlled temperature and humidity, but with many industrial facilities, environment can be unpredictable. For a window manufacturer's facility, for example, heavy volumes of fine sawdust could easily coat camera housings and get into system components. With large facilities it may not be practical to heat or cool buildings to anything much different from the outside temperature. In other facilities, such as steel mills, temperatures can reach unbelievable levels that would literally melt a camera.

The extreme environmental conditions that may occur in manufacturing and industrial facilities are also a big reason that camera systems are needed. Although there may still be a concern about theft in this atmosphere, worker and plant safety is often the driving force behind having a surveillance system installed. Large dangerous machinery, extreme temperatures, products and equipment stacked two stories high, and the presence of dangerous chemicals are all valid reasons that a plant may wish to install video surveillance. These conditions, along with the fact that many such facilities operate 24 hours a day, make system design for manufacturing and industrial environments one of the most challenging tasks a system designer could face.

Manufacturing Design Considerations

When designing a system for a manufacturing or industrial facility, a few things are unique compared with other facilities. First, most businesses install camera systems to help watch for loss of products or information as the primary function. In many cases, in an industrial environment, however, the primary purpose of the system is to help watch for safety violations or loss of life.

Taking a safety approach to system designing creates a type of system different from that most security experts are familiar with. Cameras in a loss prevention system, for example, focus on entrances and exits as a primary target, whereas industrial systems focus primarily on internal danger areas of a facility. Determining the biggest areas of concern can take full cooperation from many departments of the business. Including all departments may also help justify the camera system and share the costs among different divisions. Loss of products and information are also still concerns as a secondary purpose of the system, so some traditional coverage areas will be used as well.

As mentioned earlier, one of the biggest design considerations may be the environmental conditions within and around the facility. If a particular area of concern is that the building continuously operates at 285°F, for example, then a standard camera and enclosure will probably not hold up sufficiently. A water-cooled or oil-cooled ballistic housing may be required to keep the camera temperature low enough to perform properly. Cabling to and from the high-temperature areas would also be a concern. It would be important to make sure that any cables and connectors that are used are well protected and rated for use in such environments.

If cameras will be installed in areas with high levels of dust, smoke, or airborne particles, it may be important to use camera housings with air filters or sealed, pressurized housings to protect the cameras. An environment such as this would also mean that the glass on the enclosure would need to be cleaned frequently. In this case, a camera housing with a wiper might be beneficial.

The most important thing to remember during the equipment selection phase is to find out what the conditions will be like at each camera location throughout the year. Whereas most indoor environments will remain fairly constant throughout the year, conditions in an industrial facility are much more likely to change with the outdoor climate. If the system designer does not account for seasonal changes during the design process, system-wide component failures could render the system completely useless.

Recording requirements can vary greatly from one facility to the next. Variables include purpose of the system, hours of operation, who will monitor the system, and how the system will be monitored. If the primary purpose of the system is life safety, after-hours recording can be kept to a minimum or eliminated entirely. If the facility is a 24-hour operation with three 8-hour shifts, however, there really is no such thing as after hours. If the primary purpose of the system is loss prevention, after-hours recording should be done, preferably with video motion detection and alarm triggers to enhance recording capabilities. The best recording capabilities would still be required during the operational hours of the business.

For 24-hour facilities, the best way to record would be one tape per recorder per shift. Each shift would be required to change the tape, producing an excellent frame rate and an excellent way to archive the video. Tapes could be labeled with the date, start time, recorder number, and shift number to make finding archived video in the future much easier.

Location of the control equipment is an important part of design in the industrial environment. Most equipment, such as multiplexers and recorders, is not designed to withstand harsh conditions, so it must be set up in a more controlled environment. Most such facilities have an office area, guard station, or command center that is ideal for control equipment. The best location is somewhere that has controlled temperature and humidity levels. Because of the other equipment installed, that is usually the case with command centers and data centers, which makes them the best choice for camera equipment installation. Many facilities attempt to install the camera control equipment in a shift supervisor's office. These offices are usually in or near the main working area and are often prone to drastic environmental changes. The control equipment may operate properly for a time, but the life expectancy of the equipment will probably be drastically reduced.

To better understand why this situation exists, consider two factors that have an impact. If the area is prone to dust or other airborne particles, they can easily cling to either the videotapes or the recorder's recording heads. This puts an abrasive substance between the tape and the head every time the unit is recording or playing back tapes. This can pit the recording heads and wear out the recorder and the tapes prematurely. Gradually, the recording quality will decline until the recorder must be replaced.

Temperature, particularly high temperatures, can also have a negative impact on the equipment. Most equipment undergoes life-cycle testing in an environmental laboratory at some point. This burn-in testing occurs when a piece of equipment is operated near the top end of its operational temperature limit for a period of usually 48 to 72 hours.

The effect on the equipment components is the same as if the equipment was operating at normal room temperature for a much longer period. High temperature decreases the life span of the individual components within a piece of equipment; therefore, the life span of that piece of equipment is decreased. If the equipment is operating near its high-temperature limit during day-to-day operations, it cannot be expected to survive for nearly as long as a unit operating in normal office conditions.

Another design consideration for industrial facilities is the use of cameras for monitoring process control equipment, such as temperature indicators, tank-level indicators, or environmental gauges. By setting a camera to view the gauge or device, it can easily be monitored remotely, keeping people out of harm's way. This may not necessarily be a security function, but it can help justify a camera system or increase the life safety of the employees, potentially reducing a company's liability.

CAMERAS IN THE GAMING INDUSTRY

To see some of the most elaborate and state-of-the-art camera systems in the world, a person needs only to wander the floors of a Las Vegas or Atlantic City casino. In fact, gaming industry video surveillance has been a driving force behind some of the greatest innovations in the video industry. Any given casino can have millions of dollars passing through on any given day, even any given hour. Because so much money is at stake, gaming facilities attract every person who thinks he or she has a way to beat the system, usually by cheating or stealing. Dishonest gamblers and dealers alike could have a profound impact on a casino if they were not caught. From counting cards to stealing chips or sliding the dice, someone attempts to cheat or steal from casinos and gamblers on a daily basis. Counting cards may not be deemed as cheating, but it is still frowned on by casinos because it reduces their odds of coming out ahead.

For casinos, a high-quality, well-designed camera system can pay for itself by capturing just one incident. An effective camera system equates directly to an increased profit margin. Because millions of dollars can be won or lost on a single wager, the surveillance budget is not nearly as much of a concern as it may be with other types of businesses. In fact, the more that is invested in the camera system, the more is usually recovered by exposing cheating attempts.

Another key factor with gaming facilities is the fact that many dishonest patrons view casinos as bottomless money pits. Because of this attitude, casinos are prone to false claims for slip and fall injuries, sliding door injuries, faulty equipment incidents, and anything else that is difficult to disprove. Without adequate video surveillance, a casino may be forced to pay a large settlement for something that 30 seconds of videotape could have had thrown out of court.

Most people involved in casino security do not like to discuss the intricate details of a casino's video surveillance system, and for good reason. If all of the details of casino security were public knowledge, many people would be better equipped to find ways to get around the security. The fact is that if they beat the camera system, they beat the casino. Therefore, few specific details about casino security will be discussed in this book. Any examples given are merely generalizations that do not reflect the practices of any one specific casino. This section will discuss a few generalities that will not affect the integrity of any facility. The main reason for any mention of the gaming industry at all is the fact that gaming requirements and design considerations can have a tremendous effect on new technology and other industries.

Casino Design Considerations

Casinos are one of the few establishments for which specific regulations have been developed concerning video surveillance. The State of Nevada has implemented such regulations to help casinos in their ability to combat dishonest gamblers, dealers, and patrons.

One common characteristic of camera systems in gaming establishments is a high number of cameras per facility. It is not uncommon for one camera system to include more than 600 cameras, all being continuously monitored and recorded. Each table and slot machine is covered in most cases by at least one camera, and often more. Color cameras are mostly all that will be used on a casino floor. Because chips are small and distinguished only by color variations and numeric markings, it can be difficult to discern one value from another as well as to pick out counterfeit chips. With monochrome cameras, such minute details would be virtually impossible to identify.

Although fixed cameras and lenses are used as well, many systems have a high percentage of pan/tilt and zoom cameras, often in speed domes. A speed dome can zoom in as much as 128 times and have the ability to pan and tilt at speeds much quicker than traditional pan/tilt units. This capability can be important when the simple flick of a wrist can mean the difference between a legitimate win and someone who is cheating.

With traditional camera systems in most businesses, live monitoring and playback of recorded video is done only by trained security professionals. A trained security staff also monitors casino camera systems, but the dealers monitor the activities of gamblers and other dealers. A trained dealer can identify cheating techniques much more effectively than someone trained in security measures. Knowing that other dealers are monitoring them also helps keep the dealers honest. Security staff will still monitor the camera activity for traditional security problems, such as a rowdy drunk, people stealing chips, and false injury claims.

In summary, it should be obvious at this point that the purpose of a camera system can vary greatly from one facility to the next. It should also be obvious that designing a camera system is not as simple as picking a few products out of a vendor's catalog and installing them together. This approach may work to create a camera system, but it would be sheer luck if that system was efficient, effective, and met the needs of the facility.

The next few chapters cover specific information about various components of a camera system without getting too technical. By combining the information about these components and how they fit into the system, a security manager should be better equipped to make an informed decision about system design. Whether the manager's job is to create the design, approve the design, or use the camera system, the next few chapters should provide the information necessary to do so more effectively.

This chapter could continue on for many pages talking about every possible type of facility. The point, however, is that the purpose of the camera system should be specific to each individual site. A detailed analysis must be performed in advance if an effective surveillance system is to be designed. The samples covered here are but a few and are meant to provide insight and ideas for determining system requirements. Many facilities are actually combinations of many different facility types, so several of the examples given may apply to just one building. A large retail store, for example, may also have a large warehouse and loading area that would need cameras designed for an industrial environment. The store will probably also have office environments for the support staff, accounting service, and other departments. It may even have a day care area for employees and parents who are shopping. Trying to install a cookie-cutter camera system in such a facility will inevitably leave many people dissatisfied with the system, the designer, and probably the installation company. If a survey is done in advance and the true purpose of the system for each section is determined, an effective camera system can be designed and installed.

2

Determining Necessary Requirements

ADVANCED PLANNING

When designing a camera system, the most important part of the procedure takes place well before any system component is selected. Regardless of the size of the facility or the size of the system to be installed, each design is unique. Each should be considered as more than a combination of cameras, recorders, and monitors and should be designed to meet the specific present and future needs of that facility. To do this properly, these needs must first be determined. The best way to determine the facility's needs is with an in-depth interview of the facility occupants, careful review of the blueprints, and, most important, a close look at a risk analysis or threat assessment performed to evaluate areas of concern (see Figure 2.1).

THREAT ASSESSMENT

Although it is not the purpose of this book to teach the techniques of performing a threat assessment, the importance of performing one when designing a camera system must be mentioned. To fully understand what will be expected from the system, we must first determine what possible events could occur and the critical impact of those events. This assessment does not always need to be as in-depth as a full facility risk analysis, but certain types of occurrences at camera locations could affect the way the camera should be set up and recorded. For example, in a warehouse environment, a forklift accident is highly possible, whereas an armed robbery is less likely to occur. In a convenience store, a camera at the front counter would need to be set for the possibility of an armed robbery, whereas a forklift accident is highly unlikely.

Figure 2.1 Camera location can depend on many factors, including esthetics.

The act of determining these possibilities and their likelihood of occurring is called a *threat assessment* or *risk analysis*. Camera systems designed simply by facility layout, budgetary requirements, or compiling the newest components available have little chance of satisfying the owner's perceptions of performance if they are not based on the owner's true needs.

As with any threat assessment, three steps are required to define the needs of a camera system:

1. Determine the types of threats to assets that may occur. This includes any possible threats to people, property (including information), and privacy associated with the facility.

2. Determine the likelihood of any of these threats occurring.

3. Determine what impact each of these threats will incur on the business and personnel should they take place.

By carefully evaluating each risk possibility and the potential impact on the organization, several key factors for the camera system can more accurately be determined. In areas with the greatest risk of loss or highest impact on the organization, coverage can be increased to reduce the risk or at least record the loss for preservation as evidence. By using cost justification in conjunction with the threat assessment results, we can determine how much money is justifiable to spend on a system. This in turn can help us determine the size of the system to be installed. System size will also rely on the priority elements of the system. For example, it may be determined that a multiplexer and an additional recorder are more important than four more cameras.

System size and budget should also be based on the loss history of the business, among other things. If the loss history is in the thousands of dollars and a risk analysis shows no reasonable threat of higher loss levels in the future, a camera system costing more than $100,000 is hardly justifiable. This example may seem extreme, but excessively sized systems are sold and installed regularly.

The criticality of each possible occurrence should also be emphasized during any cost justification, such as the scenario mentioned previously. If a similar situation showed a slim chance of a child being taken from the area, such as in a play area, then the impact on the business would be tremendous if that should happen. The likelihood of the event happening may be small, but if the impact is high enough to potentially threaten the business, it must be addressed.

Designing a system based on the potential threats to a facility may seem like common sense, but that is not how most systems are designed. In fact, most systems are designed using guesswork and available items. In most cases, the company desiring a system will determine how much money it is willing to spend, usually based not on loss concerns but on annual budgets. The annual budget is important, but it should not be the deciding factor for the system design. The system should be designed first, then the budget applied to purchase essential components. The system should be designed so that additional items that did not fit into the budget can be added to the system using funding from later years.

In most cases, once company personnel determine how much they are willing to spend, they contact an installation company, which assigns a sales representative. The salesperson then determines what components will be used in the system based on the budgetary restrictions and the equipment that the installation company sells. This is not to say that all installation companies or salespeople design systems this way, but because of their training and motivations, key elements may be missed.

It is recommended, therefore, that only someone trained in performing threat assessments and all aspects of system design undertake this process. In addition, the details of the system layout should be discussed extensively with the requesting organization before anything is finalized. Care should be taken to ensure that the manager's expectations and input are included in this design process regardless of who performs the assessment and design.

EXPANDABILITY

One key element of a camera system that is often overlooked is expandability. A company looking at a camera system will often design it so that the number of cameras is equal or close to the total number possible. For example, a multiplexer might be chosen first with 16 camera inputs. A multiplexer is a camera control device that provides the ability to record up to 16 cameras on a videotape and to view up to 16 cameras on a single monitor. The system layout would be designed next, choosing the best 16 locations for cameras. Suppose that a system is designed with 12 cameras and the designer selects a multiplexer with up to 16 video inputs. In some cases, the four vacant inputs may seem like a sufficient amount of expandability, but if five additional cameras are ever needed, the multiplexer may need to be replaced. If this unit could not be connected to other multiplexers and the user wanted to go beyond 16 cameras, he or she would be faced with three options. He or she could replace the unit with another one with more video inputs that could be bussed. If a multiplexer is capable of being bused, it means that multiple units can be linked together to expand the total system. If each multiplexer can handle 16 cameras and the system can handle 16 multiplexers, then the system can accommodate 256 cameras.

The user could also install a second multiplexer for the additional cameras and have two camera systems that do not interact. Or the user could forget about the additional cameras and stop at the maximum of 16. If the system is designed properly from the beginning, the designer can plan for expansion based on the threat assessment, company growth history, and comparison of the system to those in similar environments.

A few components in any advanced CCTV system can be expandable. Matrix switchers, which act as a hub to distribute video from multiple cameras to multiple monitors and multiple control stations, are one item that has several expandable components. Common environments for matrix switchers are facilities in which multiple groups of people, such as guard forces and managers, must be capable of viewing live video images from multiple cameras in different

Figure 2.2 Matrix units are available in a wide range of sizes. This Digiplex system by Kalatel is expandable to 512 camera inputs and 64 monitor outputs.

areas of the building. These areas may include guard posts, reception areas, and managers' offices. Several portions of the matrix switcher are usually expandable, and all should be examined closely when designing the system (see Figure 2.2).

The first portion of a matrix switcher that is available in varying sizes would be the video inputs. Smaller units with a fixed number of inputs may have 16 to 32 video inputs. Larger systems usually have a mainframe and a combination of circuit cards that can be inserted to customize the system to the number of inputs desired. With larger units, up to 512 cameras can be distributed throughout the same system, and with some units 1,024 inputs are possible (see Figure 2.3).

Other areas of the matrix switcher that can vary in size are the monitor outputs and the number of control keyboards. The monitor outputs determine how many locations the various cameras can be viewed from, and the control key-

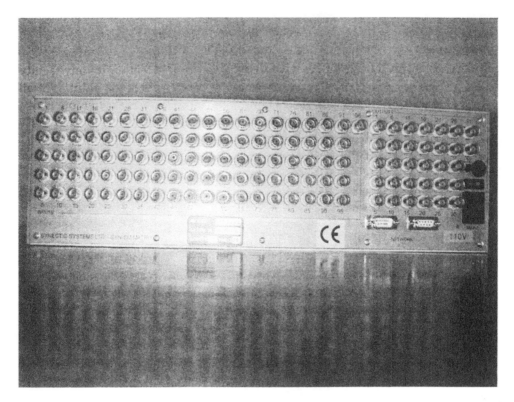

Figure 2.3   While some matrix units have expansion board slots to increase the system capabilities, others do not. This matrix unit by Synectic Systems has 96 camera inputs and 32 monitor outputs. Expansion is accomplished by networking multiple units together.

boards determine how many stations the displays and cameras can be controlled from. Alarm inputs and relay outputs are also available in variable configurations and should be closely analyzed before determining which unit to acquire.

Multiplexers are camera control units that take multiple camera inputs and allow the user to view, record, and/or play back any combination of cameras at one time. Typically, the user can view from 1 to 16 cameras on one monitor at the same time, with some units capable of displaying 32 cameras simultaneously. The most common reason for adding a multiplexer to the system is for the ability to record multiple cameras onto a single tape. This feature greatly reduces the number of recorders required, compared with recording each camera and compared with recording the output of a switcher.

Expandability is an important consideration when choosing a multiplexer. Most manufacturers only allow for 16 video inputs and 2 video outputs. Many manufacturers, however, allow for the interconnection of several multiplexers, usually up to 16, with some type of network or bus. This allows for the control of several multiplexers from a single unit or from multiple locations, giving the user access to as many as 256 cameras. This is an excellent feature for any facility that may expand the camera system beyond 16 cameras at some point. Not all multiplexers are designed with this expandability, so if there is any remote possibility of expansion beyond 16 cameras, it is important to plan for this eventuality in advance. It is far more cost effective to purchase a unit with this feature during the initial installation and add a second unit when needed than it is to replace the unit later and buy one more for expansion.

When the system is being planned, matrix switchers and multiplexers are often chosen that can only handle the number of video inputs required at that time. What usually happens, however, is that the user gradually finds areas where additional cameras would be helpful, the needs of the facility change, or the facility may even be expanded. Larger units are usually not considered initially because they are perceived as too expensive. With matrix switchers, however, the larger units are usually modular, meaning that they can be configured to meet the current needs at a lower cost. Adding additional inputs or outputs later is quick and relatively inexpensive.

The cost of expanding a system designed just to meet the initial requirements is usually far more than adding on to a system designed with expansion in mind. In many cases, the initial cost difference is minimal and the long-term savings in equipment and labor can be phenomenal if expansion is figured into the initial design.

REDUNDANCY

Another key element to designing a camera system is to include some levels of redundancy. By having additional cabling installed and backup components available, system downtime and service costs can be greatly reduced as the system ages.

After a system has been installed for several years, cameras, cables, connectors, and control equipment have an increased possibility of failure. If some level of component failure is anticipated, the designer can prepare for it and design some backups into the system to keep the downtime and repair costs to a minimum. That is not to say that the designer should double every cable run and control item, but in some crucial areas, such as fiber-optic backbones and keyboard control cables, the initial cost of duplication can be far more cost effective than future repairs.

Although it may not be possible to have an entire matrix switcher as a backup, adding one unused video input card with eight open inputs means a future card failure can be overcome in minutes. For cable runs as a redundancy, alternate routing throughout the facility is an excellent idea when possible. The biggest advantage is that if something should occur to cut or damage a cable, alternate routing means that the backup cable will more than likely be unharmed. Changing the connections over to a second cable will take a minimal amount of time compared with repairing or replacing the damaged cable, which would also be much more expensive than the cost of installing the second cable in the first place. From a security aspect, alternate routing makes it much more difficult for someone to compromise the system.

An important point to consider when determining which cable runs to duplicate is accessibility and impact on the system if the cable is damaged. The initial costs of running additional cables are usually minimal compared with the potential cost of service calls, repairs, and replacement if it is not done. Bill Norman, retired Security Director for the New Zealand Embassy in Washington, DC, sums it up by saying: "It is a lot cheaper to pull two cables once than it is to pull one cable twice." Labor costs are usually much lower for installations than for service calls, and the amount of labor required is lower when other cables are being pulled anyway.

Monitors, cameras, multiplexers, and recorders should also be considered as possibilities for backup units. Although the multiplexers and recorders do not seem like a feasible possibility to include for a backup, there is a justifiable way to include this plan into the initial design. With many systems, the video inputs are connected through a multiplexer and recorded onto a time-lapse recorder. The tapes can then be played back through the same system, but if the multiplexer is not equipped with an additional recorder, the activity going on while reviewing a tape will not be recorded.

One way to avoid this potential void in recording time is to include a second multiplexer and recorder used solely for playing back these tapes. If an additional unit is used for playback only, it will usually have far less usage than the main unit. This means that the life expectancy of the playback unit will be substantially longer than that of the main unit, making the playback unit ideal as a backup. If the primary multiplexer or recorder ever fails, the playback unit could be temporarily used until the primary unit is either repaired or replaced.

Monitors are a relatively low-cost item to add for a backup unit. If monitors are rotated so the image is not always the same, picture burn-in can be avoided and the life expectancy of each can be greatly increased. Monitors also

Figure 2.4   Keeping spare cameras on hand can eliminate downtime in the event of a camera failure.

have a longer life expectancy with a darker viewing area and little or no bright, contrasting areas.

Cameras are also a relatively low-cost item when compared with the larger control units. By keeping an extra camera or two readily available, downtime can be greatly reduced, avoiding the possibility of missing crucial events. A good rule of thumb to use if possible is to keep a 10 percent supply of cameras available. For example, systems with 16 cameras should keep two spares available, and systems with 32 cameras should keep three on hand. This may seem like overkill with larger systems, such as those with 250 cameras, but an organization that has a system that large will probably have cameras spread throughout several facilities or a very large area, so it may be necessary to maintain a supply in several different areas (see Figure 2.4).

This observation is based on my personal experiences and is only a suggested spare parts level. Many larger systems that are located in a single facility would not require nearly so many spares with a good maintenance program. It is a good idea, however, to have at least a minimal amount of equipment readily available to avoid any system downtime.

It is important to learn exactly how manufacturers fulfill their warranty requirements. Some manufacturers provide advance replacements for broken units, but others may require that the defective unit be shipped back for repair and returned after repairs are complete. The exact policy could affect the number of spares required on hand.

Key Elements of the System

This chapter describes the various functions that are performed with a camera system and the important factors of each for the overall design. Each function may seem simplistic and self-explanatory, but each has important characteristics that must be considered for the best overall system performance.

VIEWING

The most basic element of any camera system is the ability to view remote areas covered by the cameras. Setup and design for viewing becomes most important when a system will be continuously monitored, such as a casino, retail loss prevention operations, and facilities with a full-time guard force.

If a system layout is not designed properly, the personnel may have difficulty seeing activity accurately, defeating the purpose of the system and the monitoring personnel. Common mistakes are using monitors that are too small or too large, monitors that are positioned too close or too far away, improper height, and/or too many cameras per monitor for clear viewing.

Viewing of activity on a camera system is usually done with closed-circuit monitors designed specifically for security cameras. Not all systems use this type of monitor exclusively, however. Viewing cameras through standard computer monitors has become more widespread as computer-based systems become more readily available. Camera management software and video capture cards also make it possible to modify older systems to accommodate integration into computer-based security functions. When computer viewing is used, however, it is often not from the primary viewing area but from the workspace of a system manager or security manager. Both types of viewing setups will be discussed, but first and foremost will be traditional closed-circuit monitor applications.

Figure 3.1   This setup in a retail facility allows the user to view various sections of the building at a glance and is divided up for easy navigation through nearly 100 cameras.

Monitor size and arrangement is an essential part of designing any guard station or reception area with CCTV capabilities. Unfortunately, with many facilities, the monitor layout is an afterthought and not designed to fit into the environment. Often, any available monitor is chosen, with very little consideration given to how and where it will be installed, ventilation and cleaning it may require, or how much usable space will be taken up (see Figure 3.1).

The size of the monitor will vary, depending on the installation. The smallest monitor commonly used is a nine-inch viewing area. A monitor's stated size is the distance diagonally across the viewing area, such as from the top left of the screen to the bottom right of the screen. This is the same dimension used when describing the size of a television.

A nine-inch monitor is adequate for viewing a single camera view at a close range, especially if the viewing station will require several monitors. For a quad view, or viewing multiple cameras on the same screen, a larger monitor should be used. With four images that small, it becomes much more difficult to notice any activity that may be of importance.

For viewing of multiple cameras, a 12-inch monitor should be used at a minimum, and a 14-inch monitor would be preferred. With a 14-inch monitor viewing four cameras simultaneously, each camera would have 7 inches of viewing area, only slightly smaller than a separate 9-inch monitor.

Arrangement of the monitors is also an important part of the design. With too many monitors spread out over several feet, it becomes impossible for a single person to sufficiently view all cameras. For a single person, monitors should be installed at eye level from 24 to 36 inches away. The monitors should be in the direct line of sight of the operator so that peripheral vision is not necessary. If equipment is required to control the cameras, such as a multiplexer or pan/tilt and zoom controller, a monitor for viewing the changes made should be in direct proximity to the control equipment (see Figure 3.2).

An excellent example of proper monitor layout can be found in most casinos. Casinos use a high volume of cameras, monitors, and monitoring personnel to efficiently observe the entire facility. Because of the large number of cameras used, the gaming industry has established guidelines that ensure one person is not watching too many monitors and can adequately see all cameras under command.

For smaller facilities and those without full-time monitoring personnel, there is a little more flexibility for setup. For example, if a camera is being used to see when somebody needs to be given access to a door, the monitor could be placed at a reception desk. It could also be wall or ceiling mounted and angled toward the person who must grant access. If a wall or ceiling mount is used from any distance, a larger monitor, such as 12 to 14 inches, should be used to make sure the person on the other side of the door is identifiable. Because only one specific action is being looked for, it is much easier to identify, even with a smaller monitor.

In the smaller facilities, a large monitor may be used to view all of the cameras simultaneously, if the system has that capability. Most multiplexers will provide a split screen with up to 16 cameras on screen at the same time. A good application for this is a security manager's office, where he or she checks the cameras periodically but does not watch all of the time. This setup can also work in a full-time monitoring situation if the personnel are properly trained in viewing and system operation. For a person to adequately monitor a large number of cameras from one monitor, he or she must be able to identify any noteworthy activity and quickly change view to enlarge the image with that activity.

In addition to the size and location of the monitors, several important factors determine the proper view from each camera. The type of camera and loca-

Figure 3.2   Layout of the control room is a very important element of the overall system design. Monitor size and location are among the important considerations when establishing the control room layout.

tions, type of lens, and its size all play a role in the image quality and a person's ability to efficiently monitor any activity. Each of these factors is discussed further throughout this book, but two overall concepts must be discussed and understood here first.

One frequent mistake made when laying out a camera system is improperly established viewing areas. It is usually assumed that to identify a person, the image should be as close as possible. It is also usually assumed that to properly cover an area, a wide-angle view is necessary. Neither of these assumptions is necessarily accurate, and each situation should be looked at more closely. If a camera is used to cover a specific area and to identify people or objects in that area, a few guidelines can help us set up the viewing area.

Before the viewing area can be determined, the desired viewing subject must be identified adequately, which is usually part of the initial site survey or site analysis. If the system design is being done during a building's blueprint stage, this can be a little more difficult and must be based on an area's dimensions and hypothetical situations.

Viewing Area

For an example of establishing a proper viewing area with a particular camera, we will look at the front entrance of an office building. The main purpose of a camera in this location would be to monitor the traffic flow of people in and out of the building, in order to identify unauthorized personnel in the area, theft of objects through that entrance, or someone causing a problem in that area.

Common sense says that if the camera is set to look at the entrance doors, anyone coming or going could be viewed and caught on tape. Although this solution may be adequate in some cases, there are some inherent flaws when making the door the viewing subject. The most important flaw is that while a camera inside of the facility looking at the door will capture everyone coming into the facility, it will only show people from the back when they are leaving. The biggest problem with this scenario is that if someone is stealing something, the object may be seen or concealed from the camera, but either way the person usually could not be identified beyond reasonable doubt from the back. Another flaw with this setup is the potential lighting problem.

If the camera is looking at the doors during the day and it is brighter outside than inside, people entering would have a lot of backlight (more light behind them than in front of them). This condition would make a person appear as a darker object on screen than the background, concealing identifying features. This problem can be overcome by using cameras with backlight compensation, which is discussed more in later chapters. When possible, this situation should be avoided.

A better approach would be to further analyze the viewing subject. Although the doors should be part of the subject, they should not be the sole focus. Instead, look at the traffic patterns of people entering and leaving through the doors. Usually, people will be coming from or going to one of the two particular areas and will travel along the same path. If those paths, as well as the doors, can be viewed on the screen from the side or an angle—from 45 degrees to perpendicular—much more information can be obtained from the images.

Figure 3.3 The viewing area could vary depending on the purpose of the camera in each location. In this camera view, for example, the viewing area could be set as just the cash registers or the entire customer area, depending on the concerns.

The doors are now on the far left or far right of the screen, and backlight from outside is much less of an issue. Whether people are coming or going, they will be viewed from the side and are much more accurately identified. Direction of travel can now be determined, and the person can more easily be tracked throughout the facility by picking him or her up with other cameras. If a person is leaving with a stolen object, even if the object is concealed, the area he or she came from can be more easily identified, where another camera may have captured the object in question.

If multiple paths can be taken to or from the doors, more than one camera may be necessary to adequately cover the area. Trying to cover too much with one camera means that a little from each path may be covered, but nothing will be covered efficiently.

Now that the viewing subject has been determined, the viewing area must be established (see Figures 3.3 and 3.4). In the earlier example, the viewing subject was the door and the traffic paths leading to and from the door. The viewing area will be the actual limits—from left to right and top to bottom—that are dis-

Figure 3.4 Recording video images can be a key factor of the camera system. If camera images are going to be recorded, it is important to evaluate how they will be recorded and played back.

played on screen of that viewing subject. Although the actual viewing area will vary by the size of the monitor and the type of activity that will be viewed, the viewing subject should not be too large or too small within the screen. The viewing subject in any picture should take up no less than 40 percent and no more than 75 percent of the total display area. The actual size will vary in each situation, depending on the type of activity being monitored, number of cameras within the same coverage area, size of the viewing monitors, and the amount of detail required within the viewing area.

If the viewing subject takes up less than 40 percent of the screen, it will be too small to capture necessary details, and vital information may be lost. In this situation, the camera is trying to cover too much at once. Either the viewing area should be made smaller or the viewing subject should be reevaluated and multiple cameras should be used. On the other extreme, if the viewing subject takes up too much of the screen, the image may be too close to catch important details. Also, filling the display with the viewing subject does not allow for variations to the traffic patterns or abnormal circumstances, such as a second party outside of the viewing area, which changes the interpretation of the display.

The 40 to 75 percent rule may work well where a fixed camera is required, but it will not work in every situation. For many retail applications, for example, the viewing subject on some cameras may change from moment to moment. Although it would be possible to cover all potential viewing subjects with multiple fixed cameras, many times it is not practical, either because of cost or the number of monitors to be viewed by one person and recorded.

In this case, it may be better to use a camera with which the viewing area can be changed by panning the camera left or right, tilting it up or down, or zooming the image in or out. The pan/tilt and zoom (PTZ) camera can be an excellent enhancement to the fixed cameras used for coverage. It is important to understand the proper ways to view this type of setup, but the controls and equipment required are detailed in later chapters.

For an example of proper viewing with a camera capable of changing the viewing area, a retail application is used. In a typical retail store, products are divided and displayed in various groups to make them easily located by customers. One section, for example, could be music, which is possibly further divided into compact disks and cassette tapes. The initial viewing subject in this example could be the entire music section or a row or two of merchandise. By setting up the subject (the two rows) to fill less than 75 percent of the viewing area, it will be easy to see customers entering or leaving the area under surveillance. If a person in the area raises suspicion, the viewing subject could change to a more focused area of the music section, such as compact disks. By changing to the more focused viewing area, it could be seen that the person is merely choosing the items he or she wishes to purchase, or it could possibly show further suspicious activity. In the latter case, the viewing subject could now become the suspicious person. By setting the camera so the person takes up 60 percent of the viewing area, identifying features can be captured while still viewing the activity of the person's hands and the immediate surrounding area.

If the picture is zoomed in too closely, such as to capture the person's face, the actual act of concealing an item could be missed. The good face shot could be used later to identify that person, but in order to show evidence of the shoplifting, a shot that shows the identifiable person and the act of concealment would be much preferred.

A common mistake made by untrained security personnel is to zoom in too closely and then attempt to stay with that shot and follow the person throughout the store. If the subject remains at 40 to 75 percent of the viewing area, it is easier for the security person to track the suspicious person's activities. If the person is walking toward or away from the camera, the camera can quickly be zoomed in or out to maintain the same subject-to-area ratio.

This is not to say that there are not times when a closer look is required. It may be important to zoom in closely to capture a good identity of the person or an object he or she may be holding, such as a weapon or an item he or she is attempting to steal. If the camera is used and set up properly, it can be zoomed in to catch the necessary details, then zoomed back out to the 40 to 75 percent viewing area. By doing this and maintaining a constant image of the suspicious person—or at least a portion of him or her—a proper chain of events can be established for use as evidence.

If, however, the suspicious person is out of view even momentarily, there is no way of proving that the person is still in possession of the items. The full requirements for use as evidence may vary slightly from jurisdiction to jurisdiction, so it is recommended that the security staff be familiar with the preferred methods in their immediate area.

This same possibility of a changing viewing subject can be found in many other applications, such as a camera covering a parking lot or one covering a warehouse. An important point to consider, however, is that unless someone is monitoring the camera system and can change the viewing area when needed, the PTZ camera provides limited benefit. If the system is not monitored by a person in real time, it may be better to have several fixed cameras with different viewing areas and perspectives recorded to tape for later review. If a PTZ camera were to be used and left to view the wrong viewing subject, potentially useful video could be lost from this camera. It would be better to have had a few fixed cameras capture partial activity each than one PTZ camera that captures no activity at all.

PTZ cameras can, however, be used effectively in this application if they are configured and utilized properly. Many PTZ cameras have the ability to assign presets and can actually return to a predetermined position if not used for a certain period. In other words, if a system user was moving the camera and forgot to return it to the standard viewing area, the camera will return to this position after it has been inactive for a while. Presets with alarm triggers can move the camera to predetermined positions when an alarm event occurs, helping to track the activity that the user is looking for. This subject will be covered in more detail in Chapter 7, Enhancing Recording Capabilities. The choice of which type of cameras to use goes back to how the system will be used and is an important consideration in the system design.

In a facility that does not have personnel assigned to watch the cameras all of the time, monitor viewing becomes a less critical area of concern. That is to say, monitor placement and quantity are not as crucial as they would be with someone watching all of the time. Viewing subject and the viewing area, however, remain important for the basic functionality of the system. For those facilities in which the live video is not monitored, the quality of the video that is recorded becomes the top priority.

RECORDING

Recording of the camera inputs and planning of what will be captured on the storage medium are often a poorly planned portion of the system. Recording is often given no more time than choosing a quality recorder. Deciding how much is essential to record and in what circumstances, however, is an integral part of how well the system will perform and how satisfied the user will be with the system. It is not uncommon for a company to find that when reviewing the tape of a particular incident, key portions or the entire incident have been missed. This means that the camera system was essentially useless, not achieving the results it was designed to provide. Planning of the recording that will be necessary goes back to the original risk analysis.

For each area that will be covered by a camera, certain potential activities deem that camera necessary. In each of these situations, it is important to consider how much recording is required to adequately detail the activity and how much is not enough. Some situations do not require a high number of frames per second to show illegal or unauthorized activity, whereas others will require a very high frame rate to catch every detail.

An example of the first scenario would be a storage room or warehouse after normal business hours. There may be a few people who are authorized to be in those areas who could easily be identified on tape with proper camera setup. If anyone else comes into that area, it would most likely show breaking and entering, burglary or attempted burglary, or at least trespassing. The amount of video to show this would be relatively low. A frame rate of one frame per second, even up to one frame every two seconds, could adequately show this incident. Because this type of incident should be fairly infrequent, the frame rate of the recorder could be fairly low per camera, and the use of an alarm trigger or video motion detection could increase the frame rate of that camera to one frame per second. In this situation it would be feasible to record in a 72-hour mode with multiple cameras. That means that the time-lapse recorder would record 72 hours of video onto a single tape.

The first way to achieve the increased frame rate necessary is to decrease the total recording time per tape. In the earlier scenario with the storage room, a recording rate of 72 hours per tape was easily obtainable. In the situation of a bank, however, this would not provide nearly enough information in the event of an emergency. To figure out what amount of time is more advantageous, look at the schedule of a typical bank and the events likely to occur during those times. One example where a higher recording rate may be required is a camera system in a bank. Cameras covering the teller windows would need to record as much information as possible in the event of the most likely scenario, an armed robbery. In fact, a minimum of five frames per second would provide the most detail and probably from a few different camera views. There are two ways to set up the recording to achieve this goal, depending on how many total cameras the bank has, what other security measures are in place, and how well trained the staff is to react in such a situation.

In the past it was not uncommon for an individual or group to break into a bank after hours, crack the safe, and clean out the vault. As the security systems became more difficult to penetrate, this method of bank robbery began to fade. The amount of skill required and time required in the bank made this too risky a venture for the average would-be thief. The most common means of robbing banks and many other businesses is now to simply walk in during normal business hours with a weapon, demand cash, and leave with as much as possible in as little time as possible. Knowing that camera systems are in place, most of the bank robbers tend to wear some type of disguise, making it even more important to catch as many details as possible.

Looking at a few of these possible risks for a bank, it is apparent that the highest recording rate is required during hours that the bank is staffed. This would include the time immediately before and the hours that the bank is open for business when the employees are present. A standard real-time recorder or a high-density recorder coupled with a multiplexer set to record in the 18-hour mode during these hours could provide a total frame rate of 20 frames per second. If the recorder is set to record in the six-hour mode and a T-160 tape is used, the total frame rate could be 60 frames per second for an eight-hour period. After hours, the recorder could revert to the longer recording rate of 18 hours, because any after-hours incidents would not require as many images per second. Recording at 60 frames per second during hours of operation and 20 frames per second after hours, two tapes per day are required. This would provide excellent coverage of everything during a 24-hour period and the highest reasonable security level regardless of whether or not an incident has occurred (see Figure 3.5).

In most cases, this amount of recording is not necessary. Another way of achieving a high frame rate is by using alarm triggers, which will increase the recording speed only when it is necessary. A variety of triggers can be used to activate this increased recording rate, as long as it is not clearly obvious and does

Figure 3.5   If a facility requires 24-hour video recording, it is important to have a separate reviewing station to review previously recorded tapes. This would include a monitor and the same type of multiplexer and recorder that were used to record the original tape.

not draw attention to the person causing the activation. One example is foot-activated switches at each teller station that could activate a higher recording level on all cameras or a preselected group of cameras. Another common device fits into the cash drawer and causes activation when all of the bills are removed from a particular slot. Both of these devices, as well as several others, work well as undetectable triggers for banks, convenience stores, gas stations, and virtually any facility where armed robbery is a threat.

Most recorders have an alarm input that can cause the recorder to automatically change from the selected recording speed to the minimum recording speed with the highest frame rate upon alarm activation. If a multiplexer is used, the alarm inputs of the multiplexer can change and prioritize which camera shots are sent to the recorder. In this way, the system can be customized to provide exactly the cameras desired, removing any doubt about what was (or was not) caught on tape.

The point of these two examples is to show the importance of planning what the system will record. Planning ahead to estimate what types of events may occur at each camera location and their likelihood greatly enhances the functionality and usefulness of the entire system. The same planning should go into the system whether it will be monitored live or automated just to record activity for later review. If the same approach is used with a system that is monitored all of the time, the video makes excellent verification of time and date and of what the guard may or may not have seen.

PLAYBACK AND REVIEW

If a camera system is designed to record images from the cameras, eventually those recordings will need to be played back to look for proof of criminal activity or identification of people or objects. Although the playback of the recorded video is a large part of a system's functionality, little forethought is usually given to how, why, where, or when the playback will be done, and by whom. Answers to these questions as well as several others could greatly influence a system's overall design. Failure to properly address all of these questions in advance can leave system owners and users disappointed with the system and even affect whether recordings are usable as future evidence.

Review Methods

Several techniques can be used to review the recorded video. Which one will work best for any given company depends greatly on the industry, the purpose for having the system, and the sophistication of the system. The first review method is to review the tapes only after a reported incident. There are several applications in which this method will work well and many in which it will not. In a commercial application, for example, the camera system may be installed primarily to watch for large incidents, such as violence in the workplace or theft. Everybody recorded on any given tape will probably be an employee or visitor, and there may not be anything that is extremely vulnerable to theft. In this case, it is probably not necessary to watch the detailed activities of each individual recorded.

In a retail environment, however, there is a great potential to have more unreported incidents than reported incidents. The same is true of large warehouse or manufacturing facilities, where small stack reductions of five or ten items can go unnoticed or be written off as miscounts.

If a facility is using this method of review, videotapes should be cycled and stored to provide at least one week of archived video. In most cases, any incident will have been reported in that amount of time or forgotten. In many situations, facilities may wish to maintain 30 days of archived video, however. Because many companies maintain their books, inventory, and division reports based on monthly compilation, some incidents could go unnoticed for that long. For example, a manufacturing facility may take the items manufactured minus items shipped to verify items on hand on a monthly basis. Missing items could go unnoticed until this comparison is made. Although the chance of finding the incident among one month of recordings is slim, the chance of finding it if only a one-week archive is maintained is even slimmer.

With incident-only reviewing, the amount of time required each week to review tapes can vary greatly, which will ultimately affect the answers to where and by whom the tapes will be reviewed. This potential variance is one reason the playback must be considered during a system's design phase.

Another method for reviewing the recorded images is to conduct a total review of everything recorded. This method usually requires an extensive amount of time to conduct and is not used in many applications. There are some situations, however, where total review can be a necessity. If a facility has a fast-paced environment, high volumes of people and expensive items, and easy possibilities for concealment, this method may be necessary. In retail environments, this is usually not necessary, because cash registers are totaled out for each cashier shift, and shoplifting can be detected in other ways. A company that receives, configures, and ships hundreds or thousands of computers per day may find total review necessary just as an additional check and balance. Casinos and facilities where sleight of hand could result in loss are also prime examples of places that could benefit from total review.

If total review will be done with a system, it is important to plan in advance for the additional time and equipment that is required. Reviewing a single videotape in a recorder's fastest play mode will usually require at least 2.5 hours, provided there are not too many time periods that must be played again at a slower speed. Although multiple cameras can be reviewed simultaneously, no more than eight should be done at one time, and four at a time is preferred. The more cameras that are reviewed simultaneously, the more difficult it becomes to adequately view all activity from each. It becomes easy to focus on one or two cameras and miss activity on the others, defeating the purpose of total review. Eye fatigue will also be much greater when watching multiple cameras, which applies to playback as well as live viewing.

Total review should be done using a separate set of equipment dedicated to just playing back the video. It is also best to perform the review in a quiet location with no distractions, separated from the area where live viewing is done. This will eliminate any confusion that may occur from having live and recorded video of the same scene possibly going at the same time. The chance of accidentally stopping live recording or erasing previously recorded events is also reduced.

If the recording is triggered by an alarm input or video motion detection, other facilities such as offices and retail may wish to perform total review. Because recording is event triggered, this review will be only of activity and not as monotonous as reviewing when everything is recorded. This variation is similar to event-only review and is the most recommended playback technique.

By performing playback in a separate area, there can also be a clear division of security tasks and a means of keeping those responsible for live patrols and live viewing in check. Because recorders and multiplexers in the viewing area will be used just for recording and not reviewing, they can be locked into record mode and restricted from tampering. This means that once a tape is inserted to record, the system operators cannot eject a tape or stop the recording process to intentionally miss a time period or camera view. Essentially, a higher level of recording integrity can be maintained by ensuring that the security staff cannot defeat the system. A dishonest employee may still find a way to avoid detection, but one more opportunity is basically eliminated.

If total review is to be done, playback and review will need to be done on a daily basis. The amount of video that must be archived is reduced in this case, because most images are reviewed within 24 hours. It is still recommended that a minimum of one week of archived video is maintained, however, and many companies may still wish to keep a month or more. Incidents may go unnoticed or seem unimportant during the initial review, but a later incident or series of events may deem the footage important after all. For example, a particular employee may be observed working late every night and taking home his or her laptop computer to work at home. Three weeks later, however, when the employee cannot be located, it is discovered that he or she was actually taking home a different computer each night and has vanished with 20 computers. Although one week of archiving will still show some evidence, the full chain of activity will no longer be available and the likelihood of finding footage that is admissible as evidence may also be reduced. The bottom line in many cases seems to be: When in doubt, save the tapes.

Random review of stored video is the final review technique used by many operations. This is the least effective method of monitoring a facility and is usually used when event reports are not maintained or the security group is small, understaffed, or has a low budget. The camera system is usually not as well planned and has few extras when this technique is used. These facilities are often also those most in need of adequate monitoring and security measures.

If random review is done in addition to event-based review, however, it can greatly increase the integrity of the system. By occasionally checking recordings at random, previously unnoticed activity may become evident, deeming additional review necessary. This method should be used merely as a system check and not relied on as a means of catching improper activity or behavior. When done, some methodology should be used to show that this is a regular routine and not harassment of a particular person or group. For example, a policy that states "one tape per week or month will be reviewed in its entirety and selected at random" will show that noted occurrences were not premeditated or targeted. The amount of time that video will be archived should also be noted in the guidelines, as well as a detailed description of the primary review technique to be used.

An additional angle to the question of how tapes will be reviewed is how it physically will be done. It should be determined based on the technique used whether review will be done using the same equipment as the recording or if a separate playback station will be used.

The importance of a separate playback station is much greater when the review process is extensive. Planning of additional equipment for playback should be done during the initial design, because the equipment choice and location can be so important.

The equipment required for the playback station includes a monitor, a recorder of the same manufacturer and model number as the recording equipment, and, if a multiplexer is used for recording, an additional unit of the same manufacturer and model number as the recording unit.

One possible variation to this setup is that if a duplex multiplexer is used for recording and live camera viewing, a simplex multiplexer could be used for playback. The major differences between the two will be explained in Chapter 5, Control Equipment, but essentially the simplex unit is usually substantially lower priced.

It is important that the same model of recorder and multiplexer be used, because each manufacturer and model of both can have variations in how the images are recorded. It is not always possible to view a tape recorded on one type of unit with a different type. Many multiplexer manufacturers are now stating that tapes recorded with equipment by other manufacturers can be viewed with their machines, but the result could be less than perfect. Maintaining consistency with the equipment will always provide the best possible results, regardless of which manufacturer is chosen (see Figure 3.6).

Figure 3.6   For the best results, consistency in the manufacturers of equipment used should be observed.

Which Review Methods to Use?

This question applies mainly to those systems using event-triggered recording and operations using incident-triggered review. The answer to this question should be determined during the design phase and should be as detailed as possible.

Looking first at a system that will use event-triggered recording, an extensive list of why the tapes will be reviewed can make selection of triggering devices and configuration of the system much easier and more accurate. In a bank, for example, a tape may be reviewed to study evidence of a robbery or a robbery attempt. Knowing this, what type of trigger to use to activate the recording process becomes much clearer. Because notification of an attempted robbery must be physically activated by a person, devices such as door contacts and motion detectors cannot be used accurately. A more likely trigger would be some type of pushbutton or a device that triggers when a certain bill is removed from a cash drawer.

Another example is a facility with a restricted access area that requires recording of all activity. In this type of application, a device that must be manually triggered by an operator would be inefficient and impractical. Devices such as door contacts, motion detectors, photobeams, and pressure mats could be used effectively to trigger recording of all activity in the designated area. Video motion detection can also be used effectively to ensure that recording continues as long as there is movement within the restricted area. Previously recorded video may also be reviewed as a result of certain incidents or types of incidents that may occur in a facility. Determining these potential incident types during the system design phase can help ensure that the system will meet all of the facility's future needs.

In a retail environment, a person attempting to use an altered check to make a purchase is a typical incident type that would require tape review for evidence. Awareness in advance that this type of incident would need to be captured in great detail would have a dramatic influence on the end-user's choice of cameras, lenses, viewing area, and frame rate of the recording. If the frame rate is too slow or the lens chosen gives too wide a viewing angle, there could potentially be inadequate material recorded to use the tape as evidence later. For live viewing by operators, this means that the incident could even be missed entirely as it occurs. In another example, a report of a slip and fall in a hotel could be a future liability. If there is adequate camera coverage to review the incident and determine the cause or show that the fall was faked, a costly liability could potentially be avoided. This type of incident—if noted as a possible reason why a tape would be reviewed—could affect camera placement and coverage area during the design process.

Review Locations

The decision about where recorded video will be reviewed is especially affected by whether a separate playback station is required and whether tape review will require room in the monitoring area or a separate office. The amount of control equipment used and the staff size using the monitoring and review area must then be evaluated to determine the number of desks and equipment racks, the amount of monitors, and even the floor space required. Where the video will be reviewed can also affect how control equipment will be configured and programmed. If a separate area can be used for monitoring, different security and access levels to control functions becomes much more feasible. The amount of training required can be reduced and ease of use increased for those who do not need to know how to review and play back the tapes. Lower-level system operators can be limited in the amount of equipment interaction they must have to simply change monitor views or move PTZ cameras. With less interaction, there is also a decreased likelihood of missing activity in the live viewing and less opportunity to improperly record video or alter the system programming.

Review Times

Although it may appear that this issue has already been addressed, along with how tapes will be reviewed, it actually should refer to the time and/or days that the tapes will be reviewed. This decision depends heavily on the type of review that will be used with the system. Incident-based review will usually require far less time than total review of all video. Total review, on the other hand, could require enough time each day to be a full-time job, depending on the number of cameras. The amount of time required for review could affect how recording is done. If review time is only one to two hours per day, or as low as a few hours per week, the impact of review on the security department is minimal. If review of video consistently requires more than one person and/or the major portion of the day, some method of reducing the total amount of video actually recorded is needed. There are many ways to reduce the recorded video without reducing the frames per second of necessary recordings. In fact, in most cases, the reduction in recording amount can allow for increased frames per second of the notable video. By enhancing the system to record only when something is happening, hours of tape of empty rooms and hallways can be eliminated.

Knowing in advance who will conduct the actual tape reviews can help determine the programming required and the complexity of the control equipment. If tape review will be limited to managers or a designated individual, the amount of training required for those responsible for live viewing will be reduced. This would also imply that different access levels to the system functions are desired for different job levels. It is important to establish this structure before the control equipment is chosen to decide how to achieve this goal. Because many control components are capable of multiple access levels, it must be decided which units will provide these different levels and to whom, as well as how many levels of access are required. Some typical access levels would be dayshift operator, nightshift or after-hours operator, shift supervisor, security manager, systems manager, and installer. Although it may not be required to have this many distinct access groups, a minimum of three is recommended to avoid system tampering.

Review Personnel

Operator access is primarily for those responsible for monitoring live video images. This level can be set up to allow the operator to view the live cameras, change the views provided on the monitors, and control any cameras with PTZ functions. Operators could be restricted from any recording or playback functions, limited to viewing some cameras and not others, and restricted from changing the views on monitors if desired. A typical operator access possibility would be a receptionist or guard posted at the front desk. In this location it may be desirable to allow the person to view a few of the cameras that allow verification of a person's identity and grant or deny access accordingly. A camera outside of a front door, delivery door, and warehouse door may be important for the person's job function, whereas cameras set to view activity in the core of the building may be nonessential. In this case, the operator could be allowed to view any or all of the three essential cameras, yet restricted from viewing any others.

For more advanced operator access, more system functions may be required, depending on the operator's job function. For example, for a large facility, where a guard force is used, the operator may be responsible for complete system viewing as well as alarm response. In this case, the operator should be able to control which cameras are viewed from which monitor, exert full PTZ control of the appropriate cameras, and respond to or acknowledge cameras that have been triggered by an alarm condition. Any possible action or reaction associated with the control and viewing of live camera activity should be possible for the high-level operator. The ability to affect the recording capabilities and video playback, however, should be restricted at this level.

Control of recording capabilities and overall system programming should be restricted to the security manager level or higher to eliminate the possibility of system tampering or bypassing and to reduce the chance of accidentally altering the system program. Multiplexers, matrix switchers, PC-based control units, and recording units all can usually be configured to require the appropriate passcode and verification before many of these items can be changed. Many recorders can even be locked into the record mode so that pressing stop or any other front panel button will have no effect on how the images are recorded. Levels below the security manager cannot affect the recording but can still replace the tape when it is out and possibly even begin the recording process. Once the recording begins, however, it cannot be stopped or altered without entering the proper passcode or key press sequence.

4

Choosing the Cameras

The eyes of any camera system are the cameras themselves. This is the device that gathers the images through a lens, converts them to an analog signal, and then sends them to other devices to be distributed, recorded, and viewed.

Choosing the cameras for a system can be absolutely mind-boggling. For a security manager, the difference between any one camera and another is difficult to distinguish at times. Each camera is described as the newest, the smallest, the most reliable, and many other colorful adjectives. The fact is that most cameras work the same way with some minor specification differences. Some of these specification differences can make one camera a much better choice for some situations than other cameras. Many cameras are also designed for specific purposes and applications. Although a security manager will probably not know the differences between each make and model, knowing what a camera is capable of can make selection a much simpler process.

COLOR OR MONOCHROME

Technology today has made cameras a large part of our everyday lives. From the family with a camcorder to the large corporation with an integrated camera system, cameras are everywhere. Forty years ago, the family with its own video camera was rare. With the advent of the VCR, however, as well as vast improvement in and mass production of camcorders, it seems that everyone is preserving his or her lives with video.

While consumer products have advanced tremendously, the security industry has been slow and somewhat reluctant to change. You would not consider purchasing a black-and-white camcorder or wide-screen television, yet monochrome cameras and monitors remain a staple of the commercial CCTV

Figure 4.1 These Varifocal minidome color cameras are being prepared for installation in a retail environment.

industry. Very rare is the household today with a black-and-white television set, yet monochrome camera systems greatly outnumber color systems.

Before beginning this comparison, it is important to clarify the use of the term *monochrome* and the term *black and white*. Monochrome and black-and-white are used interchangeably by most people but are technically not exactly the same thing. A black-and-white picture is a monochrome picture, but a monochrome picture is not always defined as black and white. A monochrome image is one that shows variances in light levels or luminance with various shades but has only one color level or chrominance. The term *monochrome* is derived from this difference meaning one (mono) chrominance (chrome) level. The black-and-white image will show light variances in different shades of gray and black. A monochrome image may show various shades of gray and black, or it could be derived from a different color level, showing various shades of green, for example. Most readers may have seen this when videos are shown of night vision filming for surveillance and military applications.

There are three prominent reasons that monochrome systems still outnumber color systems. The first and foremost factor is equipment pricing. Until recently, the difference in cost between monochrome and color systems was substantial. Monochrome systems have always been significantly lower priced than their color equivalents, but with new technology and more widespread use, prices of both have decreased dramatically. The price gap between color and monochrome has narrowed considerably. The additional cost of the color systems is far lower than it has ever been and is very low compared with the increased value it provides. For facilities with existing monochrome systems, changing over to color would have been expensive and painstaking in the past. Cameras would need to be replaced, which potentially means new compatible lenses. Monitors would also need to be changed, as well as multiplexers and other devices that were unable to process the color images. Typically, a security manager would have never been able to provide cost justification for such an expense.

Now, however, as a value-added improvement, and with costs dropping so dramatically, such an upgrade is much more feasible. Color multiplexers and monitors will also adequately display monochrome images, so it is not necessary to change all of the system components over at the same time. Once the control equipment is in place to allow for color cameras, the monochrome cameras can be replaced as the budget permits (see Figure 4.1).

With new system installations, monochrome systems will still be lower priced, but the added value of the color system should far outweigh the small financial savings on the initial installation. Because the labor, cabling, pan/tilt and zoom control, recorder, and lenses will cost the same regardless of the system type, only the cameras, monitors, and multiplexers will increase the cost of the color system.

The second factor in the continued life of monochrome camera systems is the misconception that color video images and system components are inferior. With the growth in computer capabilities and advancements in video enhancement, many security practitioners also believe that color is easier to alter and therefore less usable as potential evidence. The same argument is being offered now in the comparison of traditional analog recording formats and the newer recording techniques.

Most important, because video is merely a security tool, policies and procedures should be in place to protect these tools. For example, with a well-designed chain of evidence procedure and secondary means of verification, some of these concerns are irrelevant. If the proper chain of evidence is established and maintained, the opportunity to tamper with the video is taken away. If video is tampered with, experts in the field can determine what has been changed with monochrome or color.

If video is used merely as a tool, as it should be, there should also be corroborating evidence verifying the validity of the event and/or the tape. In addition, if it is such a high-security or high-profile area that this may be a concern, the designer might consider a color system with a few monochrome cameras as a redundancy and as evidentiary verifiers.

Diehards may still insist that monochrome cameras provide a clearer picture, the third reason that black and white remains so prominent. It is true that monochrome cameras and monitors in general provide a higher picture resolution than color units, but the difference is small enough that it is difficult to distinguish with the human eye. Most color cameras have at least 330 lines of resolution, with high-resolution cameras available with 570 lines or more. Monochrome cameras, on the other hand, can range from 380 lines for a standard camera to 570 lines or more of resolution for high-resolution cameras. Although a low-end monochrome camera will have a higher resolution than a low-end color camera, there is no real difference in resolution on the high-resolution color and monochrome cameras.

Picture quality will only be as good as the lowest resolution component of the system, however, which in most cases is the time-lapse recorder. Time-lapse recorders typically range from 280 to 400 lines of resolution, which means that anything above that from the camera cannot even be viewed on the playback tape. As you can see, this makes the picture resolution typically irrelevant. Higher-resolution recorders are also available, as well as digital recorders, to maximize the effectiveness of high-resolution cameras; these are discussed in later chapters. In general, the recorder will be the lowest-resolution unit for a typical camera system.

Most camera systems are not used to see the intricate details of an object. They are used for the identification of people and objects and potentially for evidence gathering. In this case, whether we are viewing an object at 330 lines of resolution or at 570 lines of resolution, the sharpness of the image is not usually the most important detail. The details and identifiers of the subject are by far the most important factors when looking at a recording or live image from a security perspective. Color far exceeds black and white as a means of identification. For example, which of the following descriptions would any investigator rather have to help a case?

- *Description 1*: A light-skinned male with dark hair, dark pants, dark shoes, a light shirt, about 6 feet tall, carrying a dark briefcase. He was last seen getting into a light-colored Ford Ranger with dark interior. The license number was 555-1212, light plate, dark letters.

■ *Description 2*: A white male with brown hair, navy blue pants, black shoes, a cream-colored shirt, about 6 feet tall, carrying a mahogany brown briefcase. He was last seen getting into a white Ford Ranger with a maroon interior. The license number was 555-1212, white with medium blue letters.

As you can see, color images provide far better descriptors than a monochrome system. Although a monochrome system may provide a higher-resolution picture, such as to read the lettering on a small object in someone's hand, the color system provides a better description of the object, such as a small gray card with a red-and-yellow logo and black lettering.

Each system has its benefits and preferred applications, but in many situations the color system is the most appropriate. In most situations where a camera system is needed, such as the previous example, the monochrome system provides far less benefit than the color system. As evidenced in a situation, the monochrome system may show a person handing an object to a clerk perhaps, but the color system will show that it was a piece of paper, not money.

UNDERSTANDING CAMERA TYPES

When the time comes to install a camera system, it is often thought of as a simple process. After all, it is nothing more than counting how many cameras are needed, deciding on the size of the monitors, adding a multiplexer, and adding a recorder, right? This downplay of the importance of the design usually leads to dissatisfaction with the system, the recording, and the group who installed it all.

When a security manager chooses a camera to be installed, it is usually based on black and white or color, cost, availability, and esthetics. Some managers may also know a few other criteria, such as resolution or lens size, but typically the camera specifications are too confusing to factor in the decision. These factors are important aspects of equipment selection, but they should be secondary considerations after determining performance requirements.

Camera choices are essentially divided up into seven different types. Each of these camera types is available in either monochrome or black and white, or color. Within each camera type, cameras may be available as standard-resolution monochrome, high-resolution monochrome, standard-resolution color, and high-resolution color.

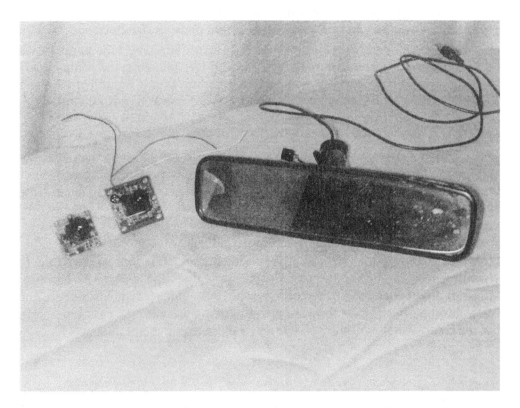

Figure 4.2   Pictured left to right: a black-and-white board camera with a 3.6mm lens, a color board camera with a 3.6mm lens, and a black-and-white board camera with a pinhole lens concealed inside an automobile mirror.

Board Cameras

A board camera is a small camera consisting of a lens mounted directly to a circuit board or small group of boards. Power and video cables usually connect to this type of camera with a wiring harness and a miniature connector or a group of small wires.

Board cameras may or may not come inside a small case to protect the circuit boards. The lens is usually fixed and not interchangeable. They are available in a wide range of lens options, although most are usually equipped with a 3.6mm lens or a pinhole lens (see Figure 4.2).

Board cameras are most frequently used as covert cameras or in areas where size and visibility are key factors. This type of camera requires 9 to 12 volts DC to operate and usually has very low power consumption. This quality makes board cameras ideal for temporary and short-term camera installations. A wireless transmitter can be used, and the camera and transmitter can be powered from a battery, the size depending on the length of time necessary to transmit the video.

Board cameras can also be used effectively for permanent camera installations in which esthetics is a big concern. Because of their size, these cameras can be installed inside numerous objects for concealment. Even if installed in plain view, they will not draw much attention.

Board cameras are available in various sizes, some smaller than one inch by one inch. There is also a range of picture qualities from standard to high resolution. They are also readily available in color or black and white.

Bullet Cameras

Bullet cameras use similar technology to the board cameras with a different configuration. Also known as lipstick cameras, these units are small cylinder-shaped cameras in a metal housing. They are available in color or black and white, standard or high resolution, and with a variety of lens options. Most bullet cameras come equipped with a 3.6mm lens for a wide-angle view, but many can be ordered with various lens sizes up to 12mm. They are also available with a varifocal lens. Most bullet cameras include their own camera mount and will not mount easily on a traditional camera mount. In most cases, a traditional camera mount is much too large because most bullet cameras are less than 4 inches long.

For outdoor applications, several small bullet cameras are weather resistant or even waterproof and submersible to 60 feet in water. Weather resistant means that the camera could be used outside but should not be directly exposed to weather conditions. This means the camera could be mounted under an overhang or something similar so that it is not continuously subjected to rain or the midday sun.

When considering bullet cameras for outdoor applications, pay particular attention to the operating temperature range in the specifications. Bullet cameras are self-contained and are not available with heaters or cooling fans, and there is no need for an environmental enclosure. The operating voltage is typically 12 volts DC, and other options are not readily available.

Bullet cameras are ideal for situations where budget and esthetics are a big concern. They are typically lower priced compared with a traditional camera and lens and are an affordable alternative for many fixed camera locations.

Figure 4.3 Bullet cameras are also known as lipstick cameras, because of the shape of the cases. Pictured are two such cameras, along with a UFO camera.

One drawback to the bullet cameras is lack of features compared to many of the more expensive traditional cameras. Most bullet cameras are equipped with built-in auto-iris control, but most do not have backlight compensation and lenses are not easily interchangeable. Camera locations that require pan/tilt and zoom control cannot use bullet cameras (see Figure 4.3).

Another important item to remember is the connection method for bullet cameras. Most come equipped with a cable and connectors in place. The power connector is usually a standard transformer plug-in. Video connectors are available as a BNC connector, which is used with most security camera systems. Most cameras are equipped with RCA connectors, however, which is the same type used on consumer video recorders and televisions. Cameras equipped with RCA connectors usually include an RCA and BNC adapter to more easily add the cameras to a traditional camera system. These adapters simply push on, with no means of locking them in place. Care should be taken to protect this connector,

Figure 4.4 Pictured are a fixed dome camera covering a specific viewing subject and a pan/tilt and zoom dome camera covering a region of a facility.

because it will be more susceptible to corrosion or degradation, particularly in outdoor applications.

Dome Cameras

Dome cameras can actually be a deceiving category including several different camera types. To understand the range of dome cameras better, it is best to look at them in two separate groups (see Figure 4.4).

Fixed Domes

A fixed dome camera means that the camera within the dome enclosure remains in one position. Camera system operators cannot change the view from that cam-

Figure 4.5 Small fixed domes are available in a wide range of styles. Pictured is a color fixed dome camera with a varifocal lens.

era in any way. Inside the dome is a camera and lens, which can be physically adjusted to the desired view and secured into place.

With dome cameras, the trend among manufacturers is to make them smaller and smaller (see Figure 4.5). Many fixed dome cameras are actually nothing more than a board camera inside a small plastic dome-shaped housing. These cameras have all of the advantages of the board cameras, such as low power consumption and multiple options; however, they also may lack some of the benefits of more expensive cameras, such as backlight compensation. One benefit that is added is the increased visibility or presence compared with the board camera. Also, unlike a standard board camera, it is more difficult for the casual observer to identify exactly what the camera is viewing. Most of the domes have a mirrored finish that acts like one-way glass. Others are mirrored except for a narrow strip directly in front of the lens. From a distance it is difficult to distinguish this strip from the mirrored portion of the dome.

Larger fixed dome cameras are also readily available for multiple applications. These larger domes usually provide enough space for a full-sized camera and lens, which allows the designer to choose any camera and features desired. Although more expensive than their smaller counterparts, these domes do not restrict the designer as much regarding available options. The dome portions for the larger domes are similar in design to the smaller domes but usually come with more options. They are usually available as mirrored plastic but can also be clear, smoked, or gold tint to fit more appropriately with the building design.

Dome size itself can vary from 6 to 20 inches depending on the application. Domes are available in indoor or outdoor configurations as pendant mount, pole mount, surface mount, or recessed enclosures. *Pendant mount* means that the camera hangs from the bottom of the mounting arm with the dome portion facing downward, similar to a pendant on a necklace. *Recessed mount* means that only the viewing dome portion is easily visible to the casual observer. This type is usually mounted in a ceiling, either in a ceiling tile location or directly into a drywall ceiling.

Outdoor models can be fully equipped with heaters and cooling fans, as well as being waterproof. The camera in the dome can be color, black and white, or even a day/night camera. Day/night cameras are color during the day and black and white at night. Outdoor fixed domes can have a wide range of mounting hardware and can be selected to look identical to the domes that have pan/tilt and zoom capabilities. Because there is no distinguishable difference, it is difficult for anyone to determine exactly where the camera is aimed. This is a distinct advantage over outdoor cameras in traditional camera housings.

PTZ Domes

Fully equipped pan/tilt and zoom (PTZ) domes provide the camera system operator with the ability to move the camera left and right (pan) or up and down (tilt). They also allow the operator to change the view on the camera with a zoom lens, closing in on smaller areas of the subject field.

To control the PTZ functions of a camera requires a receiver. This receiver converts the command signal sent through the system from the operator into the proper relay closure to perform the desired function. With PTZ domes the receiver is built directly into the dome, which means a much easier and cleaner installation. There are several types of command communications from the controller to the receiver, and most domes are able to work with more than one type.

There are several added features with the PTZ domes that can be an advantage over a traditional PTZ camera. A dome configuration can move in all directions much more rapidly than a regular pan/tilt unit. In addition, a dome can turn in a continuous 360-degree rotation and aim the camera straight down.

Figure 4.6 PTZ dome cameras allow the user to vary the viewing area and subject from that camera location. When used properly, these cameras can greatly enhance the capabilities of a camera system.

A dome camera can more easily track a person or vehicle directly below it and utilize a feature called *auto-flip*. Auto-flip causes the camera to rotate automatically when the subject passes straight below the camera, making sure that the viewing subject remains upright on the screen. This means that a subject would not go out of view while the camera is repositioning (see Figure 4.6).

Many domes also include alarm inputs directly at the camera location, as well as prepositioning. What this means for the designer is that multiple areas in a camera's potential viewing range can contain triggers to automatically reposition the camera. Even when the monitoring station is unattended, the cameras will move to areas of activity to send back the video images of what is happening in that area. Contacts on doors, motion detectors, duress buttons, and countless other items can be used to trigger a reaction by the camera. Preposition settings allow the system programmer to set up the camera view for a specific type of occurrence. If a door contact activates when the door opens, the camera could move to look at that door, and the zoom lens could set to the desired field of view automatically. Multiple motion detectors could be used to have the camera automatically track someone walking through a particular area.

These prepositions also allow the system to be programmed for automatic camera functions. In other words, even with no activity or alarm triggers, the camera could perform a sequence through multiple positions, staying with one view for as long as it is programmed to, and then moving to the next desired position. At the end of the string of views, the camera would return to a predetermined resting view. This feature may only be available when this type of camera is connected to control equipment that is also capable of providing this function.

An important consideration to remember when using PTZ domes in outdoor environments is that these units will require much more power to operate properly. The power supplies and cabling must be sized accordingly to make sure the domes receive the proper voltage and current for operation. If the power supply is not sized properly, the camera may operate properly certain times of the year but will blow fuses when heaters or cooling fans are activated. If the cooling is not sized properly for the distance required, activating the heater or cooling fan could cause the voltage to drop too low, shutting down the camera.

Full-Size Cameras

When most people mentally picture a security camera, a full-size camera is what they envision (see Figure 4.7). This type of camera is typically purchased as just a camera body. Lenses, mounting hardware, enclosures, and pan/tilt units are all

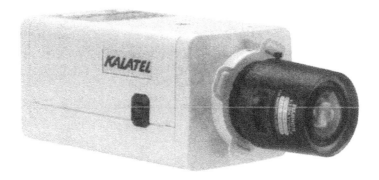

Figure 4.7 Full-size cameras such as this Kalatel unit must be properly matched with an appropriate lens. Note the black connector on the side of the camera, which allows an auto-iris lens to be used.

extra items that complete the list of equipment for a camera location. Unlike the dome cameras and bullet cameras, these units are not plug and play, with everything necessary to view a particular area.

Full-size cameras have the widest range of options to choose from. If the designer is looking at full-size cameras, understanding the specifications will be crucial to decide which one is most appropriate for a particular application. One important step when using full-size cameras is to make sure that the best possible camera and lens combination is chosen. A high-quality camera with the wrong type of lens will produce mediocre results at best. Choosing an expensive camera and an expensive lens is no guarantee that they will work well together. Therefore, it is important to understand the characteristics of both.

For outdoor applications, environmental housings must be selected carefully for compatibility with the camera and lens. This step is relatively simple, because most camera housings differ mainly by their size. A few different styles will have limits on the size of the camera and lens. Most important is to make sure that the camera, lens, and cable connections will easily fit in the housing and that the temperature range and ventilation with the housing are compatible with the camera.

As for camera mounts for full-size camera, the biggest deciding factor is the weight capacity they can hold. The designer must also pay attention to the length of a wall mount to make sure the camera has adequate room away from the wall. If the mount puts the camera too close to the wall, it may not be possible to pivot the camera to obtain the desired view.

Some of the features available with full-size cameras are not yet available with board cameras and bullet cameras. Very high resolution and very low light levels, for example, can only be obtained with some of the full-size cameras. Some specialty full-size cameras designed specifically for low light levels are capable of displaying fully detailed monochrome images with less than starlight available.

Zoom lenses are also available for full-size cameras but not for board or bullet cameras. Varifocal lenses, which act like a manual zoom lens, are most common with full-size cameras but are now available for the smaller camera types. A varifocal lens usually has a range of view, with the most common being about 3.5mm to about 8mm. This means that the camera view can be precisely adjusted without trying multiple lenses.

Network Cameras

Network cameras are the newest type of cameras in the security industry. Instead of the traditional video output from the back of the camera, this camera connects directly to a computer network. The camera is actually a self-contained video server with its own node or address on the network, as if it were a computer. The video from a network camera can be viewed from any computer on the network, provided that the computer is equipped with the proper software. Most network cameras can be viewed with a standard Web browser, the same programs used to view Web sites on the Internet. Because the cameras are self-contained servers, many can actually be connected and configured to transmit and view the images over the Internet. Many can also be accessed directly from a computer by a dial-up modem connecting straight from the viewing computer to the camera.

Network cameras are still in the early stages of use, and as such have advantages and disadvantages. One of the advantages is that special cabling is no longer required. The camera can plug into any computer jack on the local area network (LAN) or wide area network (WAN). Most facilities with a computer network will have one or more connection points in nearly every room.

One disadvantage is the ability to tie the camera into an existing camera system. Most control equipment is only equipped with connectors for a traditional camera system. Recording the images from the network camera is more complicated than simply connecting to a recorder and pressing the record button. A designer who is considering using network cameras should fully evaluate the capabilities and limitations of this system before making a decision.

UNDERSTANDING CAMERA SPECIFICATIONS

Reading and analyzing the specifications of a camera can be like trying to solve a mystery. It is not usually necessary for a security manager to be a technical expert, but understanding the specifications can be important for camera selection. The first part of camera selection to understand is the different types and categories of cameras that are available.

Every camera has a list of characteristics called the specifications. The type of information included in the specifications is fairly consistent from one manufacturer to the next. The manner in which the information is portrayed, however, can vary greatly. Because there is no set requirement for listing the information, manufacturers can display the information in the manner that shows that particular camera in the best way possible. The camera user must understand the information and know how to compare specifications of multiple cameras accurately.

Nearly all cameras operate the same way electrically, so there is no need to go into a detailed theory of operations. Although theory is an interesting topic for technically inclined people, it will not give a security manager much insight into choosing cameras or designing a functional system. For the purposes of this book, it is not as important for the reader to understand how a camera works as it is to understand how to determine how well a specific camera will perform in a particular environment. When the specifications are understood, they provide much more information and can make the camera choice simpler.

Image Sensor

Every camera has an image sensor that picks up the image and starts the conversion from light variations to an electronic signal. The camera lens gathers the light or image of the scene to be viewed and concentrates it onto the image sensor.

Most cameras in use today will utilize a charge-coupled device (CCD) type of image sensor. Some camera specifications will call this the pickup device instead of image sensor. The information listed under this heading is usually listed in one of two ways. Most cameras list the image sensor by the size of the sensor. Typically, the specification would say 1/3-inch, 1/4-inch, or 1/2-inch format. This describes the physical size of the image sensor itself. Over the years these devices have become much smaller. Many earlier cameras had a pickup device that was 2/3 inch or 1 inch. Most full-size cameras are 1/3 inch or 1/2 inch. Most board and bullet cameras are 1/4 inch or 1/3 inch. The development of these smaller image sensors has led to much smaller cameras than were possible before.

One other way that the information might be listed is the number of pixels. A typical NTSC camera in the United States would list this as 512(H) × 492(V) pixels. This is the number of rows and columns, 512 horizontal and 492 vertical, of small dots or pixels. Some cameras will list this as 251,904 pixels to make it sound like a lot, but that is actually still 512 × 492.

Camera specifications may list the picture element next, which is listed as the number of pixels. This is the same as with the cameras that include number of pixels in the pickup device specification, 512(H) × 492(V). The actual numbers should not vary for cameras in the United States and are not really that relevant for camera selection. This number does not represent the resolution of the camera but rather the number of pixels or cells that make up a horizontal and vertical row. This number really only represents the type of system that the camera is configured to work with. It would, however, be relevant if the number is different, such as for a PAL-compatible camera. This type of camera would not work properly with a U.S. system, just as NTSC will not work properly in countries that use PAL.

Scanning System

Scanning system specifications are listed as number of lines and whether interlacing is used. This will be standard from one camera to the next within each type of camera system (i.e., NTSC, PAL). NTSC systems will list this as 525 lines and will either say interlaced or 2:1 interlace.

Scanning lines should not be confused with lines of resolution. Scanning lines refer to the number of horizontal lines it takes to complete an image. Each horizontal row of pixels is scanned at the scanning frequency. If there is a problem with the scanning frequency or the number of lines, the picture could appear jittery, roll across the screen, or not show up as a usable image at all.

Interlacing refers to the way the image is put together. When the rows of pixels are scanned, they are not scanned in exact order. An interlace ratio of 2:1 means that it takes two scans to make a completed picture. Each of the scans is called a *field*. It takes two fields to make one picture, also known as a *frame*. It is important to understand the difference between fields and frames, particularly when looking at the control equipment capabilities. For control equipment, manufacturers may state the recording capability as 30 frames per second or as 60 fields per second. This can be misleading to designers, who may think they have a higher recording capability than they actually have.

Resolution

Picture resolution is the one camera specification that most system users and manufacturers seem to focus on. Although it can be an important characteristic of a camera, it is important to remember that it is only as relevant as the resolution capability of the recording and viewing equipment.

Horizontal lines define picture resolution with a standard analog camera. This is the number of lines of pixels it takes to make up a complete picture. The more horizontal lines there are, the more definition and detail there will be in a picture. For example, a resolution of 330 lines means that the picture is made up of 330 lines horizontally. A resolution of 450 lines means that the same picture is divided into 450 lines. Because there are 120 more lines to make up the same picture, the individual rows will be closer together, providing much more detail.

When choosing the cameras for a system, resolution is most important when it is used for comparison with the resolution of the monitors and recorders. It would be a waste of money to install high-resolution cameras and use average monitors and recorders. In many cases a high-resolution recorder and monitor paired with medium-resolution cameras will be more evenly matched. It is best to make sure that the resolution capabilities of all three components are fairly evenly matched.

Video Output

Video output is the description of the type of signal that is being sent from the camera through the cable, and possibly the type of connector that is used. Typically, this specification would be stated as 1-volt peak-to-peak NTSC (or PAL) 75 ohm BNC. Very rarely, if at all, will this specification show anything other than 1-volt peak-to-peak for an analog camera. This is the standard signal for an analog camera that is recognized by the viewing and control equipment.

This 1-volt peak-to-peak signal actually consists of several portions that are usually not covered in the specification. Technical readers may wish to learn more about what makes up the video signal output with a more technical book, but it is not essential at the user level to know about the various portions of this signal.

The next portion of the specification, NTSC or PAL, refers to the standard type of video system that the camera is made for. For camera systems in the United States, the specification should say NTSC, whereas in Europe the specification should say PAL. Readers who are unsure of what the standard is for their country should find out before purchasing any cameras on their own because cameras are not usually compatible with the other system type(s).

The reference to 75 ohms refers to impedance matching for the camera system. Traditional CCTV systems all use 75 ohms for impedance matching. This means that the camera must have a 75-ohm output and the cabling used must be rated as 75-ohm cable if traditional coaxial cable is used. At the end of the video line, the device that the cable is connected to must be set for a 75-ohm termination. Typically, this is seen on the rear of monitors near the camera input and/or output. The switch would typically have two positions: one that is marked 75 ohms and one that is marked Hi-Z, which is an abbreviation for high impedance. The switch would remain in this position if the video comes out of the device to go to another device. Many newer pieces of control and monitoring equipment do not have this type of switch and the impedance matching is automatic. This means that the device will determine if it is the end of the line, and, if so, will automatically include the 75-ohm termination.

The final portion of the specification in most cases will be BNC, which is an abbreviation for bayonet N-connector. This refers to the type of cable end that is required to connect a cable from the camera to other system components.

Recently, many cameras are being manufactured with an RCA-type connector instead of a BNC connector. This is the same type of connector used on most consumer audio and video products, such as televisions, VCRs, and receivers. The video output of the cameras will still be compatible with the CCTV system, but an adapter may be required to connect the camera to traditional CCTV cabling. Typically, this adapter will also be provided if the camera is being sold as a security camera.

An important note on the impedance value of 75-ohm systems is that the cable selection must be looked at carefully. Coaxial cable such as RG/59 is available in more than one configuration. Cable used with a CCTV system must be a copper center conductor and copper shield. Cable for a community antenna television (CATV) system will have an aluminum center conductor and shield. Although both are RG/59 cable, they are not interchangeable and should not be used as such.

Minimum Illumination

Minimum illumination refers to the amount of light required by the camera to provide a good picture. This specification is usually quoted as a light level of Lux or foot-candles and the lens f-number, such as f-1.4. This can be a confusing specification and an easy one for manufacturers to quote in a manner that sounds impressive but is nothing extraordinary. A typical specification might be 1.0 Lux at f-1.4, or to sound better it may be written as .1 foot-candles at f-1.4.

Monochrome cameras will typically have a much lower amount of light required for an adequate picture when compared with color cameras. Some monochrome cameras are even designed specifically to operate in low light levels, either through their component design or through enhancement with infrared illuminators. Infrared illuminators cannot be used with traditional color cameras because they typically have an infrared block built into them.

To understand the relevance of this specification, it is important to understand the difference between Lux and foot-candles. Lux is also known as lumen per square meter. The technical definition of a lumen is one candela in one radian of a solid angle. This may be somewhat confusing, but understanding this definition is not necessarily essential to interpreting the camera specifications. A foot-candle is the amount of light cast by one candle at a distance (or height) of 1 foot. One foot-candle is equal to 9.29, or roughly 10, Lux. This difference can make the specification confusing, depending on how the manufacturer lists it.

The second part of the specification is the f-number of the lens and camera at which the minimum illumination level is quoted. The f-number is actually a ratio that describes the lens iris opening. The first number identifies the focal length of the lens, and the second number is the maximum diameter of the aperture or opening, which allows the light in through the lens. An f-number of f-1.4 means that the maximum aperture diameter is four times greater than the focal length. A focal length of f-2 would be a ratio of 2:1, so the focal length is now two times greater than the diameter of the lens aperture. Typical f-numbers are f-1.4, 2, 2.8, 4, 5.6, 8, 11, 16, 22, and 32. Although this may be confusing, the biggest thing that the reader should remember when looking at this ratio is that every next higher f-number allows for only half of the light of the previous f-number. In other words, as the f-number goes up, the amount of light allowed in goes down.

This f-number also has a tremendous effect on the picture quality and directly affects the depth of field. Depth of field refers to how much of a scene is in focus at one time. This does not refer to the image from side to side across the screen but rather to objects at various distances from the camera. Depth of field is impacted by the f-stop—the focal length of the lens and the format size of the lens (1/4 inch, 1/3 inch, etc.). A wide depth of field means that objects will be in focus in a wider range from the camera lens. A narrow depth of field means that the main object may be in focus, but objects in front of or behind the object will be out of focus. Cameras and lenses with a higher f-number will have a wider depth of field than those with a low f-number. That means that with a higher f-number, more of the objects within the viewing area would be in focus. Conversely, with a lower f-number, such as at lower light levels, the depth of field will be narrower. A shorter focal length will also mean a wider depth of field, as will a smaller lens format.

S/N Ratio

Knowing the basic meaning behind these numbers, it is easy to see how this specification can be confusing and manipulated to make a camera seem better than it actually is. The depth of field and minimum illumination may be important for some camera locations, so the wrong choice could mean dissatisfaction with the system. Although a camera may list a very low level for the minimum illumination, a very low f-number could also mean that the viewing area would have a narrow depth of field. The designer or security manager should consider these variables when choosing the cameras.

S/N ratio is an abbreviation for the signal-to-noise ratio of the video output signal. This specification will vary little from one camera to the next, but it should still be understood. A typical S/N ratio for a camera is 50 decibels. This means that the video signal level is 50 decibels higher than the noise level coming out of the camera. Most modern cameras will have an S/N ratio somewhere within 40 to 50 decibels.

The biggest consideration with this specification is that the reader should understand how the difference in numbers affects the picture quality. A camera with a high ratio will have a clearer, crisper picture than a camera with a lower ratio. The difference between 40 decibels and 50 decibels may be slightly noticeable to the human eye, but differences like 48 decibels to 50 decibels are minimal and of little concern.

Operating Temperature

Operating temperature is self-explanatory and probably needs little explanation. The biggest things to remember are that for outdoor cameras inside an environmental housing, the temperature immediately around the camera must remain safely within this temperature range. Even in harsh environments, this can usually be accomplished with the addition of heating elements and cooling fans. Indoor cameras in harsh environments may also require an environmental enclosure if there is any possibility of the temperature varying outside of this rating.

Many cameras may also specify a minimum and maximum storage temperature. This refers to the range in which it is safe to store the camera with no power connections, such as replacement units or units waiting to be installed. If a camera is stored for an extended period at temperatures outside of this range, it may not be operational when the camera is eventually powered up.

Operating Humidity

Operating humidity is another specification that requires little explanation. The humidity range that is referred to is the relative humidity in the atmosphere around the camera. Harsh environments may present a problem in this area if the conditions are extreme. If the humidity were too low, then a camera would be more prone to electrostatic shock or static electricity, which could damage components inside the camera. If the humidity is too high, it can produce condensation, which can short out components or circuit boards runs and also cause a camera failure. A humidity measurement of 100 percent is absolute saturation, which would most likely cause a camera failure unless the camera is rated as waterproof and submersible. Readers should find out the annual humidity range for the camera installation location before determining which camera to use.

Operating Voltage

Operating voltage refers to the input voltage requirement to power the camera. Most cameras operate on 12 volts AC or DC, 24 volts AC or DC, 110/120 volts AC, or, in many countries on 220 volts AC. If 12-volt or 24-volt cameras are chosen, then power supplies will probably be required, at an additional expense. Whichever voltage type is selected, it is important to plan the powering schedule and verify compatibility among the cameras in advance. It is not that cameras of multiple voltages will not work together in a camera system, but for connection to power supplies and maintaining a repair stock, it is usually more efficient to stay consistent with the camera voltage selection.

DETERMINING POWER REQUIREMENTS

Many cameras are available with more than one power option to choose from. Exactly which voltage will work best for a particular application is often not considered as a primary factor, which can be a mistake. Several variables can make one voltage choice much better than another. Power cable size, cable type, and the distance from the power source must all be considered when planning for power requirements.

Most cameras operate on 12 volts DC, 24 volts AC or DC, 120 volts AC, or 220 volts AC. DC voltage is direct current, the type of power that is used with batteries. AC voltage is alternating current, the type that is used with building power circuits.

12 Volts DC

Many of the newer-style compact cameras operate from a 12-volt DC power source. Compact cameras such as bullet cameras and board cameras are often only available as 12 volts DC, so AC power is usually not an option. Larger cameras such as PTZ domes and full-size traditional cameras have 12 volts DC as an option, as well as other voltages (see Figure 4.8).

Operating the cameras from a 12-volt DC power source offers a few possibilities that are not available with AC cameras. First, 12 volts DC is a standard voltage found in many locations in which AC voltage is not readily available. Motor vehicles, for example, operate with 12 volts DC as the primary source of power.

DC Voltage

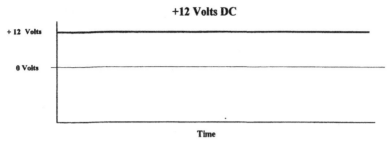

AC Voltage

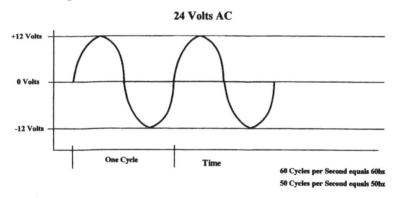

Figure 4.8 Comparison of AC and DC voltages.

Most solid-state equipment also uses 12 volts DC to operate, although usually it is created inside the equipment through power converter circuitry. This power conversion from AC to 12 volts DC is done internally on nearly all modern electronics, including AC cameras. Most of the small cameras use the 12-volt DC input directly so that the conversion circuitry is not needed. This saves on the number of components needed and the heat generated, allowing the camera to be much smaller.

Most cameras that are specified as a 12-volt DC input will usually operate properly with anywhere from 9 to 14 volts DC. This voltage range becomes much more significant when the voltage is provided through cabling of any significant distance.

Whether using 12 volts DC or 24 volts AC, voltage drop across the power cable is a concern. For the 12-volt camera, the voltage drop could be as much as 3 volts if the camera will work properly at 9 volts DC. This means that the output of the power supply would be 12 volts, but when the voltage is checked at the camera, it will have dropped to 9 volts. The 3 volts were essentially used to get the 9 volts to the camera through the cable resistance. If a bigger cable is used, the voltage drop will be less than with a smaller cable because of the resistance value of the cable itself; the smaller the cable is, the higher the resistance value will be. The higher the resistance value is, the higher the voltage drop will be across the cable.

If the voltage drop is too large, the camera may not work properly. To make sure that the voltage is adequate, the proper size power cable must be selected, and the power supply could be set to a higher output to compensate for the loss. If the power supply is set to a higher output, care should be taken to make sure it does not become too high for other cameras connected to the same supply.

Another advantage of using 12-volt DC cameras is the ability to power them from a battery. This means that if there is a power failure, the cameras can still be operational as long as any viewing, control, or recording equipment also has power. By having the ability to use batteries, 12-volt DC cameras can also easily be used in areas where power is not feasible, such as short-term construction site locations or covert installations. If a camera is to be used at a site where power is not available, it is important to calculate the amount of power required and for what period. This is best understood by understanding first how the batteries are specified. The size of a typical battery used for security equipment is stated in volts and amp-hours (AH). A 12-volt 7AH battery will supply 12 volts DC and 7 amps for 1 hour. If the equipment uses only 1 amp, it would supply the proper voltage and 1 amp for 7 hours. If the equipment uses 0.5 amp, the battery would last for 14 hours. If the camera uses only 50 milliamps (typical for many small cameras), then the battery would last for 140 hours. This means that the camera could easily be used for an extended period on just batteries, if necessary.

If 12-volt cameras are used, it is recommended that the cameras be powered from supplies specifically designed for multiple cameras. This will help ensure that all of the cameras are operating from the same ground reference, making ground loops less likely. A ground loop is when two or more cameras or items are at a different ground potential, and the difference causes a distortion of the pictures. If the voltage measured from the ground of one camera to the ground of another is a few volts, there is a ground loop. This would possibly show up on the video cable ground side, and when the video cables from the cameras are connected to the control or monitoring equipment, it causes problems with the video signal, creating distortion. Because an analog video signal is

Figure 4.9 Camera power supplies are available with individual outputs for each camera that has fuses or breakers. A power problem with one camera may not affect the other cameras if each is independently fused.

only 1 volt peak to peak, it does not take much of a difference in the ground potentials to cause a problem.

Powering the cameras from a common power source also makes future troubleshooting much simpler. Most CCTV power supplies have one individual output for each camera, each having its own on/off switch and fuse or breaker. This means that to repair or replace a camera, the power can be cut by simply flipping the proper switch. If a camera develops a problem and blows a fuse or trips a breaker, it will not affect the other cameras in the system (see Figure 4.9).

If multiple power supplies are required, it is important to make sure that all of the supplies are working off a common ground or that the cameras are connected to the system in a way to minimize ground loops. If fiber-optic cable is used, then the threat of ground loops is eliminated. If there is no way to eliminate ground differences between the power supplies, ground loop or ground isolation transformers can be used for those cameras that are causing the problem.

Most multiple-circuit power supplies are available with 4, 8, 16, 24, or 32 individually fused outputs. What is important to remember, though, is that the current rating for the power supply is the total of all of the outputs combined. This means that for a supply rated at 4 amps with eight outputs, the total is equal to 0.5 amp per output. When planning for the equipment layout, this total current capability is just as important to remember as the number of cameras connected per power supply.

One drawback to powering all of the cameras from the same source involves the voltage drop mentioned earlier. If some of the cameras are much farther away than others, they may require a much heavier gauge of wire to operate properly. This heavier-gauge wire is more expensive and more difficult to install, adding more cost for equipment and labor during the installation.

One way to address this problem is to divide the facility into zones and install power supplies closer to the camera locations. For example, if the main equipment area is on the first floor and there are six cameras on the second floor, it may be more efficient to locate a power supply on the second floor in a central location and run all of the second floor camera power from there. This means that the cable runs to each camera will be shorter, and there is one service point for power on that floor. The power supplies can then be tied to a good building ground to ensure that there is no difference in ground potential with the power supplies in other areas.

24 Volts AC

For large commercial camera systems, 24 volts AC is the most popular power source. This voltage is easy to work with, and the voltage drops for the cable runs are less significant than with the 12-volt DC supplies. With 24 volts AC, all of the cabling and connections can be done using multiple-output power supplies, and an electrician is not required to install multiple 120-volt AC receptacles.

Just as with the 12-volt DC setup, 24-volt AC power supplies are available with multiple individual outputs, making it possible to power multiple cameras from a single point. If backup power is needed, however, it is not possible by simply using batteries. Because it is an AC voltage, an uninterruptible power supply or a backup generator would be required for the cameras to operate properly in the event of a power failure. This may not be much of a problem, however, because the recording and control equipment would need the same thing. Many commercial facilities have backup power for many other things, such as the computer network, and the amount of power required for the camera system is usually relatively small in comparison to computer systems.

Most outdoor cameras and cameras with PTZ capabilities operate from either 24 volts AC or building power (110/220 volts AC), typically because they have a much higher current draw than smaller fixed cameras. This current draw would be too taxing on a 12-volt power supply, and the supply would have to be much larger. With outdoor cameras, the power supply may be required to operate not only the camera, but also a heating element and cooling fan for environmental conditions, infrared lights, or even a wiper to periodically clean the glass in front of the camera lens. For cameras equipped with these features, it is important to determine what the maximum current draw will be to ensure that the power supply can provide the necessary output. If the maximum current draw is not planned for, the picture may appear fine most of the time, but if a heater or fan kicks on, it could increase the voltage drop enough to shut down the camera. If it is close to the minimum voltage of the camera, as soon as the camera shuts down and the voltage goes back up, the camera could come back on and shut off again, causing a toggling effect. This would eventually cause a failure of the camera and/or the power supply, which would then need to be replaced.

If fiber-optic cable is used for video transmission, the designer should pay close attention to the power requirements of all equipment at the camera location. Although many manufacturers make fiber-optic transmitters that operate on 24 volts AC, some require 12 volts DC. If the camera setup uses 24 volts AC, then separate power would be required for the fiber-optic transmitter, meaning additional cabling and labor. It could be possible to include a small power supply in a box near the camera that takes a 24-volt AC input and provides a 12-volt DC output. If so, then the current requirement of this supply must be included in the total current requirement from the 24-volt AC power supply.

110/220 Volts AC

The best power source for the camera system is often the standard building power, either 110 volts or 220 volts AC, depending on the country. If the building power is used, voltage drop is no longer a concern, because the current requirement for CCTV equipment is so low. Many standard cameras come equipped with a 6-foot plug-in cable, meaning a standard wall outlet is all that is required. If this method is used, care should be taken that the receptacle is not easily accessible by passersby or someone wanting to disable the camera.

When using cameras that require 110 or 220 volts AC, there are two primary ways to install the cameras. First, as mentioned previously, a standard wall outlet can be installed at the camera location so that the camera can simply be plugged in. If the receptacle and connector are secured to prevent tampering, this method is effective. If the installation is in a more secure environment, the second option is to have the camera hard-wired into the power. This means that the power cable from the camera would run into an electrical junction box and would be permanently connected to the incoming power cable from the building. When this method is chosen, the actual connection should be made only by a licensed electrician and not by a typical security installer. An electrician will probably be used anyway to get the power cable to the location where it is needed.

One big drawback to using building power for the cameras is the increased possibility of ground loop problems. One receptacle in one section of the building could be at a different ground potential than those in another part of the building simply because they are on different circuits or different breaker panels. They may also be in different buildings or on different phases. Because the grounding for building power is fixed, the only solution to ground loops under these conditions would be installing a ground loop or isolation transformer for any camera that is causing a problem.

Another drawback to using building power is the vulnerability. Because the power is being provided directly, there is no voltage regulation mechanism in place, meaning the power is subject to fluctuations. Minor fluctuations are usually no big problem; they may just show up as a quick glitch on the screen or a slight picture roll if changing from one screen to another. Voltage spikes and surges can cause a major problem, however. If the cameras are connected directly to a building power source, they should first be routed through some type of surge protector and/or lightning protector to reduce the chance of damaging the camera.

In many cases, this is not considered during the installation, which can create big problems later. For example, with outdoor cameras mounted on the roof of a building, lightning is a big issue. If the video signal is sent through standard coaxial cable, the lightning has a good place to enter the building, causing extensive damage. Lightning always seeks the lowest ground point voltage-wise, and because the cabling puts the entire camera at the building ground, it may be the lowest ground point on the roof. This would make the camera and connectors act like a lightning rod and actually attract lightning that is in the vicinity. If the building has a good lightning suppression system on the roof, this may not be an issue. If it is an issue, however, it is best to do as much as possible to reduce the likelihood of a direct lightning strike.

Power Summary

By now it should be clear that selecting a camera is not as simple as picking any camera from a catalog or off the shelf. For each camera, many factors can directly affect which is the best choice.

Several manufacturers have taken an innovative approach to selecting the camera input power. Many cameras are now available that will operate with anywhere from 10 to 24 volts, AC or DC. This not only makes it easier to configure a system with multiple power supplies and sources, but it also makes voltage drop less of a concern (see Figure 4.10).

For a very long cable run, for example, if the power supply is 24 volts AC but only 13 volts AC is making it to the camera, the camera will still operate properly. For multiple cameras from the same power source, the length of the power cable run is, therefore, much less of a concern. A single power supply

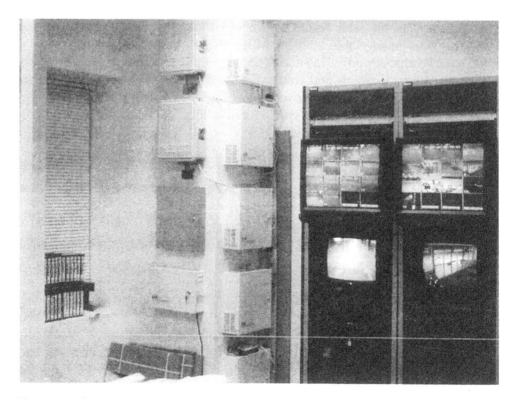

Figure 4.10 Centrally controlled power supplies for the cameras can make troubleshooting and repairs much easier, as well as reducing problems such as ground loops between cameras.

could be supplying 24 volts AC to one camera and 16 volts AC to another because of voltage drop on the cable, yet both cameras will still work properly and provide a good picture.

This ability to operate at a wide voltage range also means that the size of the power wire is less of an issue. Without this feature it might be necessary to install 14- or 16-gauge wire for longer runs, and only 18-gauge wire would be needed for shorter runs. Because the voltage drop can now safely be much larger, the 18-gauge wire can probably be used for all of the cameras.

Manufacturers of fiber-optic transmitters and receivers have also begun to make equipment that is more adaptable to multiple power sources. Although they are not set to operate through an entire voltage range like the cameras, the input voltage is selectable for many pieces of fiber-optic equipment. For example, with fiber-optic transmitters there may be three power terminals for the power input. If the power source is 12 volts, the cable could be connected to terminals one and two or two and three.

For 24 volts, the cable would be connected to terminals one and three. The input may also be selectable as either AC or DC, or with some manufacturers it could even be auto-sensing. This makes the fiber-optic modules adaptable for use with most camera locations and would require only one model of transmitter as opposed to one for each type of power. Replacement stock, inventory, and installation are then much easier, and the chance of installing the wrong type of transmitter at the wrong location is much less likely.

ADDITIONAL CAMERA FEATURES

Many camera specifications and brochures list additional features to help distinguish the advantages of that particular camera. Although basic, low-priced cameras may have few features, some cameras can have quite a few, which may be beneficial in some applications.

Automatic gain control (AGC) is one such feature that provides additional flexibility with the camera application as well as lens choice. If a camera has AGC, then a fixed-iris lens can be used as opposed to an auto-iris lens. AGC essentially adjusts the camera scene to the proper light level the same way an auto-iris lens does, but it is done electronically inside the camera itself. It increases the video signal at low light levels to make the picture brighter. Other features that work in a similar yet slightly different manner are automatic light compensation (ALC) and electronic light compensation (ELC). AGC adjusts the picture based on the overall brightness of the picture, so if the majority of a scene is dark, the AGC could adjust the entire scene to make it more easily viewable. This adjustment is done right at the video signal level, whereas an auto-iris lens actually adjusts the aperture of the lens. By using AGC instead of an auto-iris lens, the fixed lens and camera can be set up to provide the best depth of field possible.

ALC, on the other hand, acts more like backlight compensation. This essentially adjusts the entire picture to be bright enough to view the darker areas. To the human eye it appears that only the dark areas are being enhanced, but that is because the bright spots are usually already at the brightest. Scenes where a background is brighter than a foreground object would appear to be better balanced and easier for the user to distinguish objects and people. ALC is usually an adjustment directly on an auto-iris lens and not on the camera.

One problem that is often encountered is when an auto-iris lens is used with a camera utilizing AGC. If the AGC is on and the auto-iris lens is being used, certain light conditions can cause the lens and the camera to compete with each other. In other words, the lens adjusts the aperture to provide a better picture. The camera sees this change and uses the AGC to compensate for the change. This causes the lens to readjust the aperture. The lens and camera essentially toggle the light level of the picture back and forth, giving the user the illusion that the picture is pulsing. To eliminate this problem, the user can either turn off the AGC (usually done by a switch on the rear of the camera) or change to a fixed-iris lens.

Backlight Compensation

Backlight compensation is a useful feature and may be essential in some camera locations. This feature basically adjusts the image to compensate for areas that are overshadowed by bright areas. An example of where this could be beneficial is a camera looking at a front entrance door of a facility. In the daytime when it is brighter outside than inside, anyone walking through the front door would appear as just a dark spot on the screen. The light behind the person is much brighter than the light in front, and the person appears to be in a dark shadow. Without the backlight compensation, the camera or lens would adjust the view for the bright area created with the door open. With backlight compensation, the dark area in front of the person appears to be enhanced and the person, as well as the bright areas behind the person, is now distinguishable.

Digital Signal Processing

Digital signal processing (DSP) takes the concept of backlight compensation to a much more advanced level. DSP actually enhances the image digitally, which allows the camera to compensate for smaller portions of the image. Whereas backlight compensation adjusts the image based on the entire image, DSP enhances only those portions of the image that require it. The video output of cameras with DSP will still be an analog signal, but the image has been digitally enhanced to provide a much higher picture quality. All of these adjustments are done internally and require no user intervention or programming to work properly. Although these features can increase the cost of the camera, they are often worth the additional cost, and users will be much more satisfied than if they had chosen a less expensive camera.

CONCLUSION

By now the reader should realize that choosing a camera is not as simple as finding the best price and look. In fact, a single type of camera will probably not be the best choice for each camera location throughout the facility. To provide the best system possible, it may be necessary to choose several camera types for use in different environments. Decisions must also be made on the style of cameras, the use of fixed cameras, and the use of cameras with PTZ capabilities. These choices often must also be balanced with the budget allowed, and either quality, quantity, or budget must be sacrificed to find the best balance.

These same principles apply to choosing the control and viewing equipment. All have a wide price range, and often the price difference is attributable to the features provided, the reliability of the equipment, or maybe just the manufacturer. The reader's challenge is to weigh this information, decide what the priorities should be for each application, and make a decision accordingly.

5

Control Equipment

Although the main items that are carefully chosen in any camera system are the cameras, the control equipment actually makes up the backbone of the system. Many types of cameras are available for a variety of applications, but all serve the same general purpose: to send live video images from one location to another. How these images are routed, displayed, recorded, played back, enhanced, and/or restricted, however, depends on the combination of control equipment chosen (see Figure 5.1).

Camera control equipment would be defined as any device or group of devices designed to restrict, enhance, direct, preserve, or alter the video images received from a video source or sources. One of the first devices used to control a group of cameras was the video switcher.

VIDEO SWITCHERS

When camera systems started to become commonplace in the security industry, it quickly became evident that a monitor to view each camera could be costly, as well as require quite a bit of space in a viewing area. One of the first control items designed to address this problem was the video switcher. A switcher can take multiple cameras as inputs and send the video to a single display monitor, one at a time. The display seen on the monitor will change, or "switch," as directed by the switcher, from viewing the first camera to the second, the third, and so on in sequence. The switcher is also called a *sequencer* occasionally for the way these images are sequenced on the monitor. Typically, switchers are available in 4-, 8-, and 16-camera input sizes.

Although the switcher has been around for quite some time, it can still serve as an excellent enhancement to a camera system. The uses are somewhat

Figure 5.1 The Kalatel KTD-405 is a remote keyboard for multiplexer control.

limited, however, and in many cases switchers are utilized inappropriately to save on overall cost.

Switchers, when used in a camera system, serve mainly to enhance an operator's viewing capabilities when viewing live video images. It is much simpler for an operator to view eight cameras sequenced on a single monitor than it is to watch eight individual monitors with one camera each. This approach should not be used to view cameras with constant activity or in which events can take place and finish rapidly. Because the total on-screen viewing time is divided by the number of video inputs, it would be easy, even likely, that the operator would miss an incident. Cameras that could be viewed through a switcher would more likely be those covering emergency exits, those with little activity, or those used to view areas at certain times of the day. It is recommended, however, that the switcher have the ability to lock onto a certain camera in the event of an alarm trigger where applicable, such as emergency exits.

The monitor will display each camera for the same amount of time. The amount of time that each camera is displayed is known as the *dwell time* and is adjustable. It is called dwell time because it is the amount of time that each camera will dwell on the monitor screen before the next camera is displayed. Dwell time adjustments can range from a fraction of a second up to 2 minutes for each camera. For the switcher to be useful with live video images, the dwell time typically should range from 1 to 3 seconds, although this should be thoroughly evaluated for each situation.

This dwell time is also what limits the usability of a switcher for live video monitoring. Assume that an eight-input video switcher is being used with a dwell time of 1 second for each camera. Any camera connected will display on the monitor for 1 second, and then will not be displayed at all for the next 7 seconds. The amount of information displayed is only one-eighth of the information being sent to the switcher. If the output of the switcher were sent to a recorder, as often happens, it is unlikely that the recording will even be usable. One would assume that to see more information from an individual camera the dwell time could be increased. Increasing the dwell time to 3 seconds, however, means that a camera will be displayed for 3 seconds and then not displayed for the next 21 seconds. The amount of information displayed does not actually increase, regardless of the dwell time. If eight camera inputs are used, each camera will always be displayed one-eighth of the time. The ratio of viewed time to non-viewed time would remain at one to seven. The only thing that would change would be the frequency of viewing any camera (see Figure 5.2).

To demonstrate this concept, a total calculated time of 4 minutes, or 240 seconds, will be used to simplify the math. With a dwell time of 1 second per camera, it will take 8 seconds for each camera to be displayed once, 80 seconds for each to be displayed 10 times, and 240 seconds for each to be displayed 30 times. Any given camera therefore is displayed 30 times for 1 second each, a total of 30 seconds every 4 minutes. With a dwell time of 3 seconds per camera, it will take 24 seconds for all eight cameras to be displayed one time and 240 seconds to display all eight cameras 10 times. Any given camera therefore is displayed 10 times for 3 seconds each, a total of 30 seconds per camera every 4 minutes.

Capturing only 30 seconds of video from a camera every 4 minutes onto tape would have to be considered poor at best. It means that 3.5 out of 4 minutes of activity from any camera will be missed and not recorded. Unfortunately, many systems currently in place are using switchers in just this manner.

Switcher Options

When a switcher is used, a few options can help determine exactly how the unit should be connected. Many switchers have just the camera inputs and a single monitor output on the rear panel. This means that the switcher must be the last control unit in line and can be sent to just one monitor. Many times, however, the switcher will have what is known as *looping inputs* or *pass-through connectors* on the rear panel. This means that each camera will have an output as well as an input so that each camera can be passed through to another piece of control equipment or a dedicated monitor. This type of unit is preferred, because it is

Figure 5.2   Video switchers allow multiple cameras to be viewed on a single monitor in sequence.

more easily adapted to any system. With an input and an output for each camera, the unit can just as easily be the first unit in the line of control equipment as it can be the last.

Many switchers will also have a second monitor output, as well, as an option. This second monitor output is usually set to display cameras that go into alarm or a fixed individual camera that is manually set from the front panel. To clarify, a switcher may or may not have alarm inputs. If it does and an alarm trigger associated with camera three is received, for example, that camera would be displayed on the second monitor output. If a second alarm trigger associated with camera one or two were received, that new camera would be displayed on the second monitor. If, however, an alarm trigger associated with camera four or higher were received, that new image would not replace the image from camera three with most switchers unless the alarm condition was cleared already for camera three, because most switchers with this capability have the alarms prioritized from the lowest number camera input to the highest number camera input. In other words, an alarm trigger associated with camera one is the highest priority and overrides all others. The alarm for camera two overrides all but camera one, and so on up to the highest camera input number. To effectively use this feature, the importance of an alarm from each camera location must be carefully evaluated and prioritized. When set up properly, this feature can be valuable for recording, although it limits the recording to one camera input at a time and is easy for an operator to override. The most effective use of this feature is as a visual indicator to the operator that an alarm has occurred.

QUADS

As the security industry began using camera systems more frequently and began recording the video, a noticeable problem arose. Using a recorder for each individual camera input would be unreasonably cost prohibitive, and using a switcher to record multiple cameras was inefficient. The answer to this dilemma was the quad processor (see Figure 5.3).

A quad processor can take four individual camera inputs and display all four cameras on a single monitor simultaneously. The display screen is divided into four equal quadrants: top left, top right, bottom left, and bottom right. With most units each camera could also be displayed on the monitor alone, or full screen, if manually selected on the front of the quad processor. When the quad unit's output was recorded, it would essentially record exactly what was displayed on the monitor screen. If an operator accidentally or intentionally left one camera displayed full screen, the other three cameras would not be recorded at all. This made the quad, like the switcher, easy to compromise for anyone with access.

The biggest advantage that the quad offered was the reduction in the number of video recorders required to record all cameras. Eight cameras could be recorded with two quad units, eliminating the need for six of the recorders, six daily tape changes, and six of the monitors. This also greatly reduced the amount of work hours required to review tapes—a tremendous benefit in its own right. The biggest drawbacks to the quad unit were the ease of accidentally changing to a single-camera, full-screen display and the small images when viewing the quad display.

Quad processors have advanced substantially since their inception, and more modern quads have overcome the problem of recording only the image that is displayed. Many more advanced quad units are now full-duplex units, which essentially means that they contain dual processors.

Figure 5.3 Quads allow four cameras to be viewed on a single monitor at the same time, each using part of the screen. Many quads also allow for sequencing and optional configurations of the standard divided-screen view.

Operating on the same principle as multiplexers, a full-duplex quad will record the video input from all four cameras regardless of what is displayed on the monitor. This means that if a system operator accidentally sets the monitor to display just one camera full screen on the monitor, the remaining three cameras will still be recorded. This also eliminates the problem of the display of each camera being small in the quad view. If an incident is noted on a particular camera, that camera alone can be played back from the recording in full-screen mode. If the person performing the playback is unaware of any particular incident, all four cameras can be played back in the quad mode. If any suspicious activity is noted, the associated camera can be changed to a full-screen view for closer observation.

If a quad unit is to be used as part of the camera system control equipment, it is important to choose the correct unit. Features available differ greatly from unit to unit, and thus cost also varies greatly. Some options are rather easy to determine, whereas some can be greatly missed in their absence if overlooked. Many quad view units also offer passcode protection and multiple operator levels. Programming functions, such as date and time, one-screen display location, and camera titles, can be restricted from change to those with the proper authority.

If the system will have multiple parties operating it, passcode protection can be important. Also, if the control equipment is in a location that is accessible to general employees, multiple operator levels and passcode protection are essential. Although passcode protection is not very complex with most quads, it does add some semblance of protection.

Quad units are available for monochrome and color systems, and the appropriate unit for each must be selected. Color units are a bit more expensive than monochrome, but using a monochrome unit with color cameras will limit the video output to monochrome.

Although most quad units have individual video inputs and outputs for each camera, some units do have input only. In most cases, it is preferable to choose a model with individual outputs, but performance is not affected if the unit does not have this option. The cost savings of choosing a unit without individual outputs is minimal, so even if the outputs are not required for the system they are to be used with, it may be advantageous to choose a unit with them in case the configuration of the system ever changes.

Another option available with some quads is alarm inputs. An external alarm device, such as a door contact or motion detector, can be connected to the rear panel of the unit and associated with a particular camera. If an alarm event occurs, it would trigger the video output to the monitor to change from viewing a quad display to viewing the associated camera full screen. This feature is excellent if the system will be used for live viewing of camera activity. It is an instant visual notification to the operator that an event has occurred that requires immediate attention. Without this feature, if the quad display is just one view of several that the operator must observe, an incident requiring immediate attention could easily be missed. By placing the quad view monitor in proximity to the main monitors under observation, a change from the quad view to a full-screen view becomes immediately obvious to direct the operator's attention to the area of concern. A more recent variation of the traditional quad is the dual-page quad. The term *page* refers to the fact that these units have eight video inputs, which can be displayed in two groups of four.

Typically, the first screen will display cameras one through four and a second screen will display cameras five through eight. Changing the view from the first screen to the second screen can be done manually from the unit's front panel with the push of a button. Changing the view from one screen to the next can usually also be done through automatic switching with a variable dwell time. Using this method, the view will continuously change from one view to the next. This process will continue indefinitely until the unit is manually placed into another mode. Many dual-page quad units are also capable of a full-screen view of each individual camera.

Recording capabilities of a dual-page quad are typically similar to a standard quad. The recorded video will typically only be the same as what is being viewed on the monitor at the time of the recording. In other words, if the monitor is set to view cameras one through four, cameras five through eight will not be recorded. If the unit is set to cycle from the first view to the second, each set of four cameras will be recorded while on screen, but not while the other view is displayed. With the dual-page quad, it is not possible to record all eight cameras simultaneously and play back recorded video of just one particular camera.

One of the best applications for a dual-page quad is as an addition to a system that already has a multiplexer. If the eight cameras are routed through the multiplexer, they can be taken from the individual multiplexer camera outputs to the camera inputs of the quad unit. This will allow for quad view displays at secondary locations, such as a reception area, without limiting the use of the multiplexer's multiscreen view. If the multiscreen view from the multiplexer were used at the reception area and the operator's area, any changes to the display that were performed by the operator would cause the receptionist's view to change as well. These secondary monitor locations would then have a fixed display of the cameras they need with no control over system operation.

Quads are available now with several display options that make them more useful in this type of situation. The user can select full-screen view of an individual camera, quad view, sequencing, picture in picture, and variations of the standard four-segment display. If one picture is more important than the other three but all still need to be viewed, one image can be large while the other three are smaller and directly below the larger image. This flexibility ensures that the quad will be a useful component in camera systems for years to come.

MULTIPLEXERS

Multiplexers were created for the video industry to serve a dual role (see Figure 5.4). Multiplexers were primarily developed as a means of recording multiple cameras simultaneously to a single videotape. Additionally, they provide a means of viewing multiple cameras on a single monitor in various arrange-

Figure 5.4 Multiplexers have developed into one of the most widely used control devices in a CCTV system. Pictured here is the Kalatel multiplexer series, including 10-camera and 16-camera units available for monochrome and color systems.

ments. To understand the benefit of a multiplexer, it is important to understand the basic principles behind its operation.

A recorder stores video onto a videocassette in frames. Each frame is equal to one video image or picture from the input source, such as a camera. Traditional time-lapse recorders have a typical recording speed of five frames per second to record 24 hours of continuous video to a single tape. Real-time (real motion) and high-density recorders have a typical recording speed of 20 frames per second for recording 24 hours of video to a single tape.

Note: Recorders that record 20 frames per second are actually real-motion recorders. Real-time recording would require 30 frames per second. The term *real time* is used to describe a 20-frame-per-second recorder throughout this book, however, because that term is used by the recorder manufacturers. If the reader were to ask to purchase a real-motion recorder, he or she would be hard pressed to find one described as such. If, however, the reader were to ask for a real-time recorder, he or she would have dozens to choose from, all of which record 20 frames per second.

A multiplexer basically gives each camera input a single frame, dividing the total frame rate of the recorder by the number of camera inputs. The multiplexer has a recorder output that sends one image from each camera sequentially to the recorder. The speed at which each image is sent to the recorder and the space between each frame, known as the *blanking space*, must be the same as that of the selected recorder to play the images back properly later.

Looking at a multiplexer with ten camera inputs connected to a traditional time-lapse recorder, one image from each of the first five cameras would be recorded during the first second. During the next second, cameras six through ten would each have one image recorded. For the third second of recording, another image from cameras one through five will be recorded, and so on. This means that for any individual camera, the record rate will be one frame every 2 seconds. The length of time for an image is known as the *refresh rate*. An easier way to state the capabilities of this example would be to say that each camera has a refresh rate of 2 seconds.

With a multiplexer and the same ten cameras connected to a real-time or high-density recorder, the refresh rate of each camera improves drastically over the traditional time-lapse recorder. Because the real-time and high-density recorders typically record 20 frames per second for 24 hours, each camera has a refresh of 1/2 second. Each camera would have two images recorded every second. A refresh rate of 1/2 second will provide a much more accurate display of events than the 2-second refresh rate and provides a higher likelihood of recording the details of an incident.

Using the previous example of the multiplexer with ten cameras connected to a real-time recorder, the multiplexer will recompile the images at the same 20-second frame rate in which the cameras were recorded. If it were desired to play back video from camera one only in a full-screen display, images 1, 11, 21, 31, and so on would be sent to the display monitor. Images 2 through 10, 12 through 20, and so on would all be ignored by the multiplexer. If the frame rate or blanking space of the recorder and multiplexer do not match exactly, an image from a different camera or a black or distorted image could possibly be sent to the display monitor periodically. The multiplexer can usually be programmed by a technician to eliminate this problem. The biggest benefit of the multiplexer becomes much more evident when playback of recorded video must be done. Because every camera connected to the multiplexer is being recorded, the images can be played back in many different formats. Multiple cameras can be played back on screen at the same time, or any individual camera can be played back in a full-screen view. The multiplexer basically takes the video from the recorder and processes it in the reverse order of the way it was recorded.

Because of the importance of this timing, it is not possible to review a multiplexed videotape in fast forward or reverse, as with a commercial VCR and standard videotape. The images displayed on the monitor would appear jumbled and inconsistent, eliminating the possibility of honing in on one individual camera. The same is true when looking at a multiplexed videotape without the multiplexer. Every image would be displayed from every camera, and each camera would be shown too briefly to actually see details of the image.

On-screen displays can vary from 2 cameras to 16 cameras at the same time, and with some manufacturers, even 32 cameras are possible. With most multiplexers, a quad review is an option, as with a quad processor. Many 16-camera multiplexers have the ability to sequence four quad views to display all 16 cameras. Many units are also capable of customizing the multiscreen displays to meet the specific needs of the security program.

Multiscreen views from a multiplexer are often not well thought out for maximum benefit. The primary purpose of using a multiscreen display should be to group cameras with a common purpose or in a common area of concern. For example, if the multiplexer has eight camera inputs that view emergency exit doors and eight that view more active areas throughout the facility, an eight-camera multiscreen view of the emergency doors could help considerably. Because most of the activity will be coming from the other eight cameras, they may require individual monitors or a different type of grouping for the operator. The emergency doors, however, will probably have little activity, and any movement on screen would be more obvious than if several cameras had heavy activity. If the emergency doors also had an associated alarm trigger for each, any camera with an alarm could instantly be displayed on the spot monitor.

In most instances the multiscreen view of the multiplexer is thought of as no more than a means of using fewer monitors in the operator's area. Although this is definitely a benefit of using multiscreen views, it is not so in every case and should not be designed as such. If a multiscreen view is the only one used on some cameras, important information can be missed. Viewing multiple cameras on a single monitor makes the display of each camera much smaller, and for the critical live camera views in any facility, a dedicated monitor should still be used. If, for example, there are four cameras covering a common lobby or six cameras covering adjoining areas of the perimeter, having all displayed together can make an operator's job much simpler.

Simplex versus Duplex

There are two primary types of multiplexers: the simplex and the duplex. Each has distinct applications, which must be understood before choosing a unit for a system.

A simplex multiplexer can record multiple cameras, play back multiple cameras, or view multiple cameras in the live video mode. The most important point to remember about the simplex multiplexer is that it is capable of performing only one of these tasks at a time. If a simplex unit is used to view cameras in the live mode, it is not going to adequately record as described in the earlier example. If recording is attempted while in the live mode, only the images displayed on screen will actually be recorded to the tape. For live viewing, any combination of multiscreen, full-screen, or sequenced view is possible, but not without affecting the recording. If the simplex multiplexer is in the record mode, playback and changes to the live view are not possible. To accurately record all video inputs, the simplex unit must be properly set and left unchanged during the recording process. Any change to the monitor display will remove the multiplexer from the record mode and place it into the live view mode.

If a simplex multiplexer is used to play back previously recorded video, live viewing of the cameras is not possible, nor is recording of new images. This single-function-only aspect of the simplex multiplexer greatly reduces the applications for which the multiplexer can be used.

A duplex multiplexer is essentially two simplex units combined into a single multiplexer. A duplex unit can view live video, record multiple cameras, and play back previously recorded video, as with a simplex unit. The duplex unit, however, adds the capability of doing two of the three functions simultaneously. If a duplex multiplexer is being used to record, it can be used simultaneously for live viewing. Changing the display of the monitor will have no effect on the recording capabilities, regardless of which display is chosen. A duplex unit in this situation will always record full multiplexed video from all connected cameras.

Recording and playback can also be done simultaneously with the addition of a second recorder. The primary recorder would be connected to the recorder output of the multiplexer and would always be used to record new video. The secondary recorder would be connected to the multiplexer's recorder input and would always be used to play back previously recorded video. The images being played back would be viewed on the primary monitor, while the new images to be recorded would just go to the primary recorder unseen.

If the duplex multiplexer is configured with a spot monitor or a second monitor output, it might be possible to view live video images and previously recorded video images simultaneously. Depending on the multiplexer, however, this combination may be somewhat restricted. With many units, the second monitor output is only capable of sequencing the view from one camera to the next, viewing an individual camera full-screen, or viewing cameras with an alarm trigger that is in the alarm condition. With many units it is not possible to have a multiscreen view on the secondary monitor, but with some it may be an option.

With most installations that require a duplex multiplexer, the most effective way to use it is for recording and live viewing simultaneously. If continuous recording is required, it is usually best to use a second multiplexer for playback. Because playback will be the only function of the second unit, it is possible to use a simplex multiplexer to be more cost effective. One excellent design idea is to use a duplex multiplexer as the playback station. There are several advantages to using a duplex multiplexer for the playback station as opposed to the less expensive simplex unit. Because the playback station will be identical to the primary recording unit, it can be utilized as the backup recording unit if it is ever needed. Because the playback station will not typically be used as much as the primary unit, it will have a much longer life expectancy. If the primary unit ever develops a problem, the secondary unit can be used in its place until any necessary repairs have been made. Both units should be programmed the same, so downtime for a failed multiplexer will be greatly reduced. Once the primary unit is returned, it can be used as the playback station to help equalize the usage time of both units. Another distinct advantage is the ability to quickly expand or reconfigure the system if the need arises. The playback unit can be utilized as a second recording unit to add more cameras quickly when necessary. This also gives the end user the capability of installing temporary cameras for recording emergencies or unique situations.

When choosing a multiplexer, the most common mistake is to choose a simplex unit when a duplex unit is actually needed. It may seem like a fairly complex decision, because the differences between the two unit types can be somewhat confusing. The security organization is often convinced to sacrifice some functionality in order to save a little money. In most instances where a simplex unit is used for the reduced cost, the security department and those responsible for the system will soon regret it.

There is an excellent rule of thumb when choosing a multiplexer that usually will eliminate any doubt. If the purpose of the camera system is to record only, and the unit is always locked to prevent changing the function, and the system will not be used for live viewing of any associated cameras now or in the future, the simplex multiplexer will work efficiently. If any portion of that statement is not completely true, then a duplex multiplexer is recommended. If the multiplexer is to be used for live viewing only and no recording is done through a multiplexer, a simplex unit will do just fine. Carefully analyzing the system requirements in advance is crucial for the multiplexer selection, especially if a simplex unit is being considered. In most applications it will become evident during the initial analysis that a duplex multiplexer is usually the best choice.

Once a multiplexer type has been selected, it is important to review the options available for each unit to determine a specific model to be used. As with the cameras, monitors, and quads, multiplexers are available in monochrome or color. Because the cost difference can be more than $1,000 per unit, it is important to determine during the initial analysis which will be required. If any color cameras are installed or may be installed in the future, it is important to select the color multiplexer. Black-and-white cameras will still appear monochrome on a color multiplexer, but color cameras on a monochrome unit will appear black and white. In the latter case, video displayed on the monitors and recorded on the recorders will be monochrome as well, so a color camera would essentially be a waste of money. If budgeting restrictions limit the initial equipment to be installed, it would usually be better to start with a color multiplexer and black-and-white cameras and monitors than it would be to start with a monochrome multiplexer and color cameras or monitors. Cameras and monitors are much less expensive to replace individually in the future annual budgets than replacing a multiplexer.

Alarm inputs and relay outputs are portions of a multiplexer that are probably underutilized. Alarm inputs can be utilized to greatly enhance recording capabilities as well as live viewing, so they should be closely compared with each model. With many multiplexers, a single alarm input is available for each camera input, and with some units it may be less than this number. How each unit addresses the use of alarm inputs should also be considered. Alarm inputs could be through a fixed connector on the rear panel and not expandable or through alarm input modules and expansion boards. The benefit to units with alarm inputs built directly into the unit is that there are no additional components to purchase. The drawback, however, is that the amount of alarm inputs cannot usually be expanded. Units that use alarm input modules or expansion boards are usually more diverse and can allow for multiple alarm triggers per camera. The downside is that to use alarm inputs, additional equipment must be purchased. Units that have external alarm input modules usually provide much more flexibility than those with internal alarm inputs. For example, a single alarm trigger could be used to manipulate viewing and recording parameters for one or several cameras simultaneously. Multiplexers with internal alarm inputs are usually limited to one alarm input association with one camera.

Relay outputs on a multiplexer can usually be used to enhance an operator's capability and warn of any alarms or potential system problems. Lights, sirens, and other equipment can be triggered during predetermined events, such as an alarm input, to call the person's attention to that event. It is usually best to look for a unit with multiple relay outputs and flexible programming parameters for each. A single relay should be capable of being activated from several different inputs. For example, if a relay is used to turn on an alarm indicator light, it should be possible to turn on this light from alarms associated with more than just one camera. Any camera alarm or an alarm from the recorder should be capable of turning on the light to inform the operator that something with the system has happened and must be addressed for action.

Relay outputs can also be used to trigger alarm inputs on other systems, such as alarm panels or access control systems. An excellent example of an application for this option is with a recorder's "tape out" alarm. When the recording tape reaches the end on many units, a relay is activated that can be connected to an alarm input of the multiplexer. The multiplexer's alarm input can cause a relay output to be activated to turn on a warning light or sound a horn to notify the person responsible for changing the tape. Another relay output can be activated that is connected to an alarm input of an access control system. This will help provide a more traceable record of when tapes run out, when they were changed, and any possible times when no recording was taking place. Because so many systems have both alarm inputs and relay outputs, careful system planning should be done to get the most out of all systems involved. Configuration and integration of the camera system, access control system, and intrusion detection system should be thought through completely during the design and equipment selection phases.

Most multiplexers also include video motion detection as a built-in feature. This allows the user to pay more attention to those cameras that have activity within the viewing area and less attention to those with nothing going on. This feature is discussed in much more detail in Chapter 7, Enhancing Recording Capabilities.

When video motion detection is included in the multiplexer, the user can usually define which cameras use it, what areas of the picture to include, the sensitivity level, and how activity affects the viewing and recording of the images. How these options are handled will vary from one manufacturer to the next. Most, however, allow the user to mask or ignore sections of a camera view as well as reduce the sensitivity to prevent false triggers. This feature is usually underutilized and is a great way to improve the recording capabilities of the camera system.

Multiplexer Controls

Most multiplexers are easily controlled directly from the front panel of the unit. Many units, however, include an option to control the cameras remotely, allowing the multiplexer to be mounted in an equipment rack and a control device to be placed near the operator on a desk. This remote control device is usually a special remote keypad or keyboard (see Figure 5.5), which provides all of the same functions as the front panel, including programming. The control functions are often easier to perform from the remote unit than from the front panel.

Many multiplexers allow for multiple remote keyboards to work with multiple multiplexers. Users must select the multiplexer that they want to work with by the ID number and then can select cameras by the camera number. When

Figure 5.5 Remote multiplexer keyboards usually allow the user to control several multiplexers from a single device. Each unit is selected by an ID number, and then all of the multiplexer controls can be completed from the keyboard.

choosing a multiplexer, the remote capabilities and number of multiplexers that can be used together should be considered to ensure that the expansion capabilities are in line with the future needs of the facility.

MATRIX SYSTEMS

Large camera systems may have hundreds of cameras to monitor and multiple locations that need to view and control those cameras. Although it would be possible to route all of the cameras to all of the viewing locations that require them, that would not be practical. The amount of cabling needed would be tremendous.

Figure 5.6   Matrix switchers allow several users to view and control multiple cameras in several locations simultaneously.

One way to organize a camera system is with a matrix switcher. A matrix switcher will take numerous camera inputs and allow for monitoring and control from numerous locations. It provides multiple security levels, which can be assigned by operator or monitoring location (see Figure 5.6).

Matrix units for a camera system essentially act like a giant data routing network for video. They provide the ability to view and control many cameras from many locations simultaneously and can provide several security levels of access from each. Users must log on to the system from a location before they can make any system changes, change camera views, or move pan/tilt and zoom cameras.

Matrix units range in size from 16-camera inputs to thousands of camera inputs. Although smaller units come with a fixed number of camera inputs and monitor outputs, larger units are usually expandable with the addition of expansion cards. Expansion cards can add additional camera inputs, monitor outputs, control panel connections, alarm inputs, and relay outputs. Most systems that

Figure 5.7 Matrix systems can be very large, space-consuming components for a camera system. Pictured is a matrix switcher that utilizes two full-sized racks.

have expandability are based around a rack-mounted unit with a power supply that accepts the expansion cards. Cards are inserted into an available slot. These rack-mounted units are known as a *mainframe* or *card cage*. A card cage is usually equipped with a minimal number of inputs and outputs to start with. Many systems can be equipped with multiple card cages, each of which has its own unique system address (see Figure 5.7).

Control of a matrix system is usually done through a separate external keyboard (see Figure 5.8). Multiple keyboards can be used for control from multiple locations. A unique address and/or the port number on the card cage to which it is connected identify each keyboard.

Figure 5.8 Matrix units are usually controlled from a remote keyboard, similar to those used with multiplexers. With a matrix keyboard, however, there are no buttons for multiscreen views or selecting different units.

Functionally, a matrix switcher works much differently than a multiplexer and should not be considered as a substitute. Each serves a different primary purpose, and many systems will actually contain both. Matrix switchers, for example, do not provide for multiple camera views on screen simultaneously. They will, however, allow operators to bring any camera up on the monitor to which they have been allowed access. The matrix unit is a great way to supply any camera combination to any control location and restrict access. Each operator can be limited to viewing certain cameras as well as restricted by time period. User levels can be prioritized so one level has higher control than another level. When more than one operator has the ability to move a camera with pan/tilt and zoom, prioritizing can allow the higher-level user to override the lower-level user.

When a system includes a matrix unit, it should be set up as the primary piece of equipment at the control end. In other words, the video cables from all of the cameras should be routed to the matrix unit first before going to any multiplexers, monitors, recorders, or other equipment. This is primarily for more efficient layout and wiring and is not a necessity. If everything is routed to one location first, however, it makes all future maintenance, troubleshooting, and repairs easier to deal with. With the matrix serving as a distribution hub and cabling organizer, problems with either particular cameras or equipment controls are easier to isolate and troubleshoot.

Camera tours provide the ability to have a camera or group of cameras sequence through a number of predetermined positions and camera views. Any properly equipped pan/tilt and zoom camera can be set to go to a series of preset camera views in a sequence. At the end of the sequence, the next group could go through a sequence in the same manner. Going from camera to camera, the entire building or area could be covered on a single monitor view as if a guard had physically walked the same path. This camera sequence could then be set to repeat hourly, daily, or any number of times per day without operator interaction. Many matrix units have other features that are consistent with multiplexers. Although many users may want to use these features only from the multiplexer, it is often more practical and advantageous to use them from the matrix unit. Camera tours and automated camera control are features that can often be established with either the matrix switcher or the multiplexers. By establishing this feature with the matrix unit, it is more feasible to initiate or change this from multiple user locations. To use camera tours, pan/tilt and zoom cameras must also be the types that can perform the necessary functions. These cameras must be the types that allow presets, which is the ability to mark specific views as established marker points.

Alarm inputs to the matrix switcher can also be used to cause a camera to change position. An alarm contact on a door, for example, could cause a camera to move to a preset, which provides an adequate view of the door. If the alarm inputs were utilized to their maximum potential, it would be possible to track an individual through an entire facility with no human intervention.

One method to achieve subject tracking in this manner is with multiple motion detectors. Each motion detector connected to an alarm input on the matrix switcher would cause the necessary camera to move to a certain preset position. As the person or object moves from one motion detector to another, the next detector would trigger another alarm input, and the camera could change position to cover the appropriate area. If the person moves to the coverage area of a different camera, that camera could move to the necessary position for adequate coverage. This tracking method would only work properly in times when little traffic is present. In high-traffic areas the camera would be in constant motion and the video would probably not be useful.

Large camera systems present the problem of accessing the cameras quickly and easily. When multiplexers are used as the main control, the operator must usually first access the proper multiplexer and then the appropriate camera with many of the multiplexer manufacturers. Because multiplexers usually control up to 16 cameras, this can be a confusing task initially. Only the camera number identifies each camera when a matrix switcher is used. If the system contains 86 cameras, they would be numbered 1 through 86 instead of 1 through 16 and then the multiplexer number. The system operator can access any camera by simply selecting the proper camera number. Although this benefit may seem relatively minor, it is actually a big advantage, particularly if an operator must view a camera quickly.

Recording and Video Storage

Camera systems have evolved from a simple remote viewing tool to a complex and integral part of many security operations. In the past, systems were used primarily to view live activities in other areas of a building, but camera systems are now utilized to record and monitor activities throughout an operation 24 hours a day.

ANALOG RECORDING SYSTEMS

One important function of most camera systems is the ability to record, store, and play back video images in a usable manner. Traditionally, CCTV images have been recorded and saved onto videocassettes. Most typically, the video images are recorded with video time-lapse recorders (see Figure 6.1). Similar to a conventional VCR, the time-lapse recorder records the images onto the videocassette at varying frame rates.

A traditional VCR can place several hours of video onto a single tape at a rate of 30 or more frames per second—or real time to the human eye. Traditionally, the different record rates are known as slow play, super long play, and extended play. The videocassettes are also available in different lengths, usually S-120 and S-160. An S-120 tape is capable of storing 120 minutes or 2 hours if recorded in the SP mode, and S-160 tapes are capable of 160 minutes in the SP mode.

A time-lapse recorder works much the same way as the traditional VCR, but the two types are not interchangeable. With a time-lapse recorder, or VTR, the rate at which video images are recorded to the tape varies. Up to 960 hours can be placed on the tape, depending on the VTR. Typically, a VTR would have a 2-hour, 12-hour, 24-hour, and possibly 72-hour mode. The key point to remember, though, is that with anything above the 2-hour mode, not all activity from a camera will be displayed. The image on the monitor will appear to jump, like a

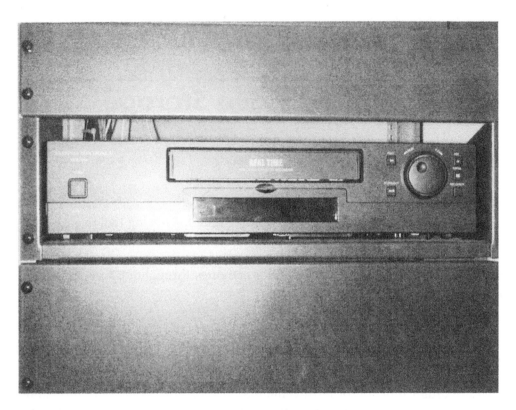

Figure 6.1 CCTV recorders come with a wide variety of options and an equally wide range of prices. Pictured here is a real-time recorder by National Electronics, which can record 20 frames per second for 24 hours onto a T-160 tape.

strobe effect, depending on the record rate. For clarification, we will look at recording in the 24-hour mode.

Most VTRs in the 24-hour mode record at a rate of five frames per second. If real-motion video is 30 frames per second, that means 5 of 30 frames are recorded and the other 25 are not. From a security perspective, that is not a major concern when dealing with only one camera, because it would be somewhat difficult to conceal activity, such as carrying something out of a building, if we view all activity five times per second. When going to a longer record rate, such as 72-hour mode, or recording multiple cameras, the frame rate becomes much more critical. In fact, frame rate and record speed alone can be the most restricting factor of a camera system. A VTR recording in the 72-hour mode has a typical frame rate of 1.5 to 1.6 frames per second. That means you would see three frames every 2 seconds. A unit recording in the 960-hour mode could record as slowly as one picture every 8 seconds. Although that type of recording is adequate for some specific applications, it is virtually useless by itself for most general security applications.

As mentioned earlier, attempting to record more than one camera through a VTR, such as with a multiplexer, will affect the frame rate per camera even more. Looking at a camera system with ten cameras connected to a multiplexer, it is easy to see why the recording rate becomes so important. The multiplexer should be set up to send video to the VTR at the same rate that the VTR is recording. As mentioned, a typical VTR in 24-hour mode will record five frames per second. Therefore, the multiplexer must send the video to the VTR at five frames per second. As discussed in the detailed section on multiplexers, video images from all ten cameras will be sent to the recorder, sequenced one after the other. Five frames per second for ten cameras means that one frame every 2 seconds will be recorded for each camera. If recording were done in the 2-hour mode at 20 frames per second, each of the ten cameras would be recorded twice per second. Recording in the 2-hour mode is not practical, however. This means that every 2 hours, the tape would have to be changed by someone. In facilities that are closed on the weekend, 24-hour mode is not practical either, unless someone will come in just to change tapes. In some instances, then, the 72-hour mode would need to be used. If the recording rate is only 1.5 frames per second, that means each camera is recorded only once every 6.66 seconds. Increasing the total number of cameras to 15 would mean that each camera is recorded only once every 10 seconds.

Obviously this is not a great way to record and store necessary security video in most applications. When attempting to record 15 cameras onto one tape for 72 hours, there is a one in ten chance of recording the exact image needed. That leaves a nine in ten chance that important information will be missed. These odds can be greatly improved by using video motion detection and alarm triggers, as discussed in a later chapter.

Another way that recording important information was improved was with the invention of the real-motion recorder. This type of recorder is more commonly referred to as a real-time recorder by most manufacturers, so to avoid confusion this book will refer to them as real-time recorders. A real-time recorder, put simply, turns the tape more slowly, recording on a smaller portion of tape. Where as a VTR recorded five frames every second in the 24-hour mode, a real-time recorder puts 20 frames per second on tape in the 24-hour mode. That means that with 15 cameras multiplexed to one tape, each camera will be recorded once every three-fourths of a second. The same situation with a VTR would produce an image from each camera once every 3 seconds. This would seem to make the choice obvious when choosing a recording unit, but most real-time recorders typically do not record more than 24 hours per tape. For retail operations that are open seven days a week, this may not pose a problem, but an office environment open five days a week is a different story. Additionally, even retail operations may be closed on holidays, leaving a situation that requires someone to come to the facility just to change the tape—or not record anything during the closed days.

If it is important to maintain the higher frame rate, but not practical to change tapes over the weekend, multiple real-time recorders could be used by taking advantage of scheduling. Unlike traditional VCRs, time-lapse and real-time recorders have an on-screen recording schedule. The video out from the multiplexer could be looped through three real-time recorders. Each recorder could then be programmed to record every third day in sequence. Although this may seem like an expensive alternative, it may be more cost effective. When the cost is broken down over the life expectancy of the recorders, and then compared with the cost of paying an employee to come in on a Saturday or Sunday over the same period, the recorders begin to seem like a practical alternative.

Many manufacturers have since improved on the real-time theory and incorporated it into a time-lapse recorder. Many models now have real-time (motion) or virtual real-time recording up to the 18- or 24-hour mode and then additional record modes of 40, 72, and/or 96 or more hours. This gives the user the option of recording at an expanded frame rate during the traditional work week, when there is more activity, and a reduced frame rate over the weekend, when there is little activity and no one available to change the tapes.

Continuous Recording

Most traditional camera systems with recording capabilities are set up to record at all times. This technique is most commonly used because it is easier to implement and requires less programming. Continuous recording can be as simple as connecting the camera or a multiplexer to the recorder and pressing the record button. There is no distinction on the tape between a recording of nothing happening and a recording of an incident. It is up to the viewer to analyze the recording on the tape to determine what is important.

One situation where continuous recording can be advantageous is when it would be difficult to distinguish between normal activity and an important incident. For example, the normal activities of a cashier in a store would show movement and interaction with customers. A dishonest cashier would show similar activity as far as electronic devices are concerned; however, a person reviewing the tape would be able to determine what activity was important. Tape review would either be performed as a periodic or scheduled review or based on suspicion of any wrongdoing.

This recording technique is effective, particularly in situations where it is difficult to distinguish between normal and abnormal activity. Unfortunately, this technique is also used far too often and, in many cases, the amount of recorded video could be drastically reduced with some planning and equipment programming.

Alarm- or Event-Triggered Recording

A camera system is often used to monitor for specific types of activities. Banks, for example, use camera systems primarily to record video in case of a robbery. Corporate offices use camera systems to record accessibility to different areas of the building and after-hour break-ins, among other activities. In either case, it is not necessary to record everything that occurs at every camera location at all times (see Figure 6.2).

Many facilities that implement a camera system can reduce the amount of video they are recording based on the type of incident they are looking for. Even retail companies can enhance the video recording capabilities to some extent by ignoring or reducing the amount of video recorded where there is no activity.

This recording technique can enhance the usability of a camera system from several different aspects. The amount of archived video can be greatly reduced, which reduces both storage and reviewing requirements. This will result in a reduction of hours required to find and/or review events, whether using a traditional or a digital recording system. The number of tapes and tape changes needed is also reduced, which means less time needed to change the tapes. This could also mean reduced wear and tear on the equipment, depending on the enhancement techniques used. The actual techniques that can be used to enhance the recording capabilities are covered in greater detail in Chapter 7, Enhancing Recording Capabilities. In most cases, these techniques will apply for analog and digital recording systems.

When alarm- and event-triggered recording is used, the frame rate of the recorder may not need to be as high as when continuous recording is used. For continuous recording with ten cameras at a total of 20 frames per second, each camera is recording 2 frames every second. With enhanced recording using a recorder capable of five frames per second, a camera with activity could use either every other frame or every frame, depending on the programming capabilities and whether other cameras have activity. Looking at a corporate office,

Figure 6.2   Areas that have no activity can have a reduced recording rate. Cash lanes for a retail store, for example, can use video motion detection to help determine when a camera view should be recorded at a higher frame rate. If there is no activity at one set of cash lanes but another set is busy, it makes sense to record the busy cash lanes more often than the lanes that are not being used.

for example, during the day there may be several cameras with activity at any given time, so a higher frame rate may be important. At 20 frames per second with ten cameras, the worst-case scenario would still be 2 frames per second per camera if all ten cameras had activity simultaneously.

If only a few cameras had activity, however, they could be recorded much more often, and the remaining frames could be divided among the cameras with no activity. The actual frame rate per camera would vary, depending on the number of cameras with activity. If only one camera had activity, it could record at ten frames per second and the other ten frames would then be divided among the other nine cameras. If two cameras had activity, they may get five frames per second each, with the remaining frames divided among the remaining eight cameras, and so forth.

This would show the best results after normal hours of operation and on the weekends, when activity within the building will be little to none. A recorder in the 72-hour mode could be used easily to cover the entire weekend. If there is any activity in the building, it would still be recorded at a high enough frame rate to be useful. Reviewing the tape would be much easier as well, because any activity will show up on screen at the higher frame rate. If the recording is done with alarm triggers only for the weekend, any alarms would be noted on the recorder, and the reviewer could then simply review the alarm video. This would eliminate the lengthy time required to sit through hours of uneventful video.

Another application where this is helpful is for camera views where activity is rare and any activity at all would be considered an event. A high-security room with limited access, for example, might require complete video coverage anytime anyone is in that room. This room may not be used often, but, when it is, the activities would need to be recorded at a decent frame rate. If the recorder were activated by a door contact and/or a motion detector, anytime anyone enters or stays in that room the system could record that camera view at all times.

Archiving Videotapes

The amount of tape to be archived depends on a few different factors. First, what length of time does the facility wish to maintain archived video for? Many companies maintain old video for anywhere from one week to one year or more. Old video is usually maintained for one week at the minimum and one month at the most.

Another factor is the recording technique that was used. If continuous recording were used, the amount of tapes required to archive for one month could require quite a bit of room. If event-triggered techniques were used, one month of video may not be so cumbersome.

The type of events that might occur can also be a factor for how long tapes are archived. Some types of incidents may not be immediately noticeable or significant until a later date. Although something like a robbery would be noticeable immediately, employee theft of a stocked item may not be detected until that item is needed at a later date. When it would be noticed could depend on the item taken and the cycle time of merchandise or equipment for the facility. A laptop computer that is shared and used infrequently might go unnoticed for days or weeks before someone realizes it is even missing. If the videotapes are only archived for one week, the video of someone leaving with the computer might be recorded over before it is noticed.

Tape-Organizing Systems

Tape-organizing systems could help track the tapes and could make it much easier to maintain archived video for longer periods. This organizing system could be something as simple as numbering the tapes and keeping them in order on a bookshelf. A log could then be kept to keep track of which tapes were used in which recorders and on which dates. The organizing system could also be more advanced, utilizing tape dispenser units and tracking the number of times each tape has been used. This would help keep track of when tapes should be replaced instead of being reused. This will help reduce the likelihood of an older tape breaking and jamming in a recorder because it was used longer than it should have been.

DIGITAL RECORDING SYSTEMS

So far, this chapter has looked at ways to record and store security video on tapes in an analog format. With the rapid advancement in computers, a new alternative has become available. By converting the video input from analog to a digital format, digital recording and storage are possible.

Technology advancements have created some major life changes in the last decade. Computer sales and development, as well as computer technology items, have grown exponentially, creating new products to replace others that were new a mere six months earlier. This rapid growth has touched virtually every industry and has created some major improvements in the security industry.

One of the fastest-growing areas of advancement in the security industry is the development of digital and computer-based video systems. In the past, computer systems could not handle the massive memory and processor requirements needed to accurately transmit, record, and store video footage properly. Video would need to be saved in short clips that would often appear jumpy and inconsistent. As the computer industry has grown, though, processors have become much quicker, hard drives have become increasingly larger, memory is larger and faster, and new storage media have been created that make digital video a viable option (see Figure 6.3).

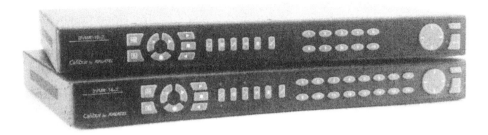

Figure 6.3 Many digital recorders are combined with a multiplexer within a single piece of equipment, such as these units by Kalatel. The DVMR 10 and 16 are capable of 10 and 16 camera inputs, respectively.

How the video is compressed and transferred has also been developed to make data retrieval much more consistent and accurate, to the point that picture quality now exceeds that of the traditional recording methods. The rapid advances in the computer industry have led to faster processors, allowing for an excellent frame rate. Hard drives have grown to allow for the massive storage requirements of a CCTV system. Digital tape storage space and speed have also grown substantially, and reliable video compression techniques have led to a reduced amount of storage space required for archiving the video.

Digital video systems are utilized for much the same reasons as the traditional camera systems. The concept of a digital system is to eliminate the common problems associated with the traditional recorded video system while increasing functionality and ease of use. There are many wonderful advantages to using a digital recording system as opposed to the traditional recording system. Picture quality, for example, is much higher with digital systems. With a traditional recording system, the recorder is usually the lowest-resolution device used. If color cameras with a resolution of 400 lines are used, but the recorder is only capable of 280 lines for color, the recorded image will only be 280 lines. With the digital system, the recorder is no longer the lowest-resolution device in most cases. This means that users will have the higher resolution they paid for when they bought the higher-resolution cameras.

Another advantage of the digital format is how well the video images hold up over time. Because each picture is digital, it is virtually a new image every time it is viewed and not as likely to degrade over time.

Standard video output is currently always in an analog format. This analog video signal must be converted to digital at some point for a digital system. Some cameras are currently available that perform analog-to-digital conversion inside the camera unit. These cameras are typically connected to a network for transmittal to a remote location either in the same facility or somewhere on the same network. They can also be transmitted for display or access via the Internet or intranets.

As stated earlier, the video signal must be converted from an analog to a digital signal before it can be digitally recorded. With most systems, the analog-to-digital conversion is done within the digital recording device, but several cameras do the conversion directly at the camera. By doing the conversion at the camera, existing network cabling can be utilized to transmit the images to the security area, where traditional cameras would require dedicated cabling throughout the facility. If the digital cameras are used, however, they must be routed to a device that can accept the digital input. Most digital video recorders still require that the video input be done the traditional way, by direct cabling and an analog video signal. There are currently four common formats for the compression and transmission of the video to a digital format used within the security industry. Each one has distinct advantages and disadvantages, and each has a niche as far as compatibility with system requirements. The four types of compression or formatting in the security industry are JPEG, MPEG, wavelet, and H-263. Each of these will be looked at more closely in the following sections. First, however, it is important to look at the actual conversion of the video signal from analog to digital to understand the differences between these different formats.

File Formats

To make an accurate comparison of the traditional system capabilities and the digital system capabilities, it is important to determine what the actual storage capacity will be for each. Many people choose to go to a digital system because of the perceived advantages and because they have heard about how great and advanced the systems are; however, they are often disappointed with the performance, because they thought it would be much more than it is. Some manufacturers may state that you can record continuously for a week, but they fail to mention that it is a very slow frame rate and/or event recording only.

The first part that is important when determining the overall storage capacity of a system is to determine the size of the files that will be recorded. With many systems the file size can be adjusted to allow for better picture quality or higher storage capacity. A larger file size will have a better picture quality but a lower overall storage capacity as far as number of frames recorded or overall recording time. A smaller file size means that more frames can be recorded, but the picture quality will be less than that of the higher file size. The type of compression used will also determine the file size that is being recorded. Each of the compression and format types will be looked at independently for a good comparison.

JPEG

If the recorder uses JPEG image types, the typical file size is usually around 15K to 31K, depending on the resolution of the images and amount of compression. Looking at a file size of 20K and a frame rate of 20 frames per second, the storage space required would be 400K per second, 24M per minute, 1.44G per hour, and 34.56G per 24-hour period. As you can see, it is not probable that recording continuously for weeks could be achieved in this situation. If the frame rate were reduced to ten frames per second, the space required would be 200K per second, 12M per minute, 720M per hour, and 17.28G per 24-hour period. If ten cameras were being recorded, that would produce one frame per second per camera.

MPEG

MPEG technology is slightly different from the JPEG, and it is difficult to determine actual recording capacity. Where the JPEG replaces the entire image with a new image every time, MPEG replaces only those pixels of an image that have changed since the last image was sent. A distinct advantage of this technology is the reduction of the file size for recording. The disadvantage, however, is that the amount of video that can be recorded depends entirely on the amount of activity on the cameras. In other words, if there are few changes to the video images, the file size could be reduced to 10K or lower. If there is a lot of activity, however, the file size will be substantially larger and will vary from one image to the next.

With the calculations of the JPEG technology, it was straightforward to determine how much could be recorded in a 24-hour period. With MPEG the frame rate would be established, but the file size for any given image is practically indeterminable. It would be possible to establish a range of the file size based on a frame with minimal activity and a frame with maximum activity. Most images are actually somewhere in between.

It would seem on the surface that having a varying file size would be an advantage if you plan for the maximum file size, and most come in substantially lower than that. Most manufacturer specifications are not written using the maximum file size to determine storage capacity, however. Most manufacturers that use the MPEG technology will state a maximum record time that will always sound much longer than the record time of systems using the other technologies. This maximum record time, however, is based on little or no activity in a camera view. Every movement and every increased file size will decrease this record time, making the maximum record time virtually unobtainable. Many users have been disappointed when they realized that the recording time they were hoping to achieve and the recording time they actually received were not close to each other.

The biggest drawback to this scenario from a security manager's perspective is the unknown of the actual recording time. If the system is set up to record 24 hours per tape, some days it may achieve half an hour more and some days it may achieve half an hour less. Although this variance may not seem that important, if nobody is available to change to a new tape once the old one is full, there could potentially be an undetermined time period where nothing is being recorded. This potential for missing crucial video footage is something that most security managers cannot tolerate with any system.

If the user understands this problem and bases the frame rate of the cameras and total recording requirements on the minimum record time, as opposed to the maximum record time, the MPEG recorder can still be beneficial to the facility. If you ask manufacturers which technology is better, they will tell you that it is the one they are using, of course. Many manufacturers and users take that a little further, though, when discussing MPEG technology if they do not use it. It is often said that the MPEG technology is less reliable than the JPEG or others. This is not because of the varying file size but because of the nature of how the file size is determined.

Because the MPEG replaces only the pixels that have changed, many JPEG users will tell you that MPEG images are actually altered video and therefore may not be admissible in court. They may argue that any given image is not a true depiction of the scene from the camera, because some pixels may be unchanged or changes too minute to warrant an update. Each pixel, however, is actually a numeric value representation of what has been sent by the camera, so if that value has not changed, there is no reason to transmit or record duplicated information.

A the time of writing this book, I am unaware of any court case in which recorded video could not be used because it was recorded with MPEG or any other technology. Although it is possible that it has happened, it has not been a major occurrence noted by the security industry.

Types of MPEG

Digital video recording for the security industry primarily uses three different types of MPEG compression. Each is similar in theory, with differences in the amount of compression and algorithms. To the end user, the only true noticeable difference will be the file size and the amount of time that can be recorded.

MPEG-1 was the first MPEG compression type to be introduced. Through continuous improvement and development, MPEG-2 was created. MPEG-2 provided a higher compression rate, meaning smaller average file sizes than MPEG-

1 images. The next progression in MPEG technology brought the MPEG-4 compression format, again with a higher compression rate and smaller sizes. Some manufacturers are using MPEG-4 technology for video storage at the time of publication of this book, whereas still more manufacturers are using MPEG-4 for video transmission and viewing, along with a different format for actual video storage. MPEG-4 allows for much easier video transmission either over a network, the Internet, or through a dial-up connection. This allows users to see more frames per second and clearer images in a shorter time frame than if they had used larger files or less compression. Other types of MPEG compression currently under development are MPEG-7 and MPEG-21, but these new formats may or may not ever come to fruition.

Wavelet

Probably the fastest-growing market share of digital recording units belongs to those utilizing wavelet technology. Wavelet technology is difficult to explain in great detail, but basically it looks at the video signal several different times and saves it in little packets of varying resolution. The first show will capture a coarse resolution, then a little finer, and a little finer until the entire image is captured. All of these packets are bundled to create one file that represents that image. Rebuilding the image is much quicker than with JPEG or MPEG, and the file sizes can be much smaller. Because of the nature of the wavelet, it can accurately compress and decompress the images much more than the other technologies.

The name *wavelet* actually gives a good hint at how this technology works, if you think about it. It looks at the entire video waveform and samples a portion of it—thus, wavelet. The use of wavelet for digital video systems has added greatly to the storage capabilities of systems and the use of digital video for remote viewing and storage. Transmitting and viewing with the other technologies require a much higher bandwidth and can be difficult on an existing network because of the load. With wavelet, however, the load is somewhat divided up, depending on the amount of network activity, allowing for more even and consistent video transmission and retrieval.

H-263

H-263 video compression is actually a standard developed for video transmission over a network. It was designed primarily to provide a usable video stream even when the available bandwidth is low. Many manufacturers are using H-263 format for network video servers and remote viewing on network-enabled digital recorders. Some use the H-263 format exclusively on the recorder, whereas others use H-263 strictly for the remote viewing function and one of the other formats, such as JPEG, for the video storage function.

The H-263 video coding standard was created to allow video streaming for lower-bandwidth applications, particularly 56 KBps (kilobytes per second). As with the MPEG standard, H-263 retransmits those portions of the video image that have changed since the previous image. The first image is used as a base image to which the next frames are compared. Those portions of the images that have changed are then transmitted, thus reducing the file sizes.

Although similar to MPEG format, the H-263 standard is not exactly the same. The algorithms that perform the compression vary from MPEG, and a few additional functions are performed with H-263. Two features that reduce file size used in H-263 are motion estimation and motion compensation. Many items within a video image are the same as previous frames except for their location within those frames. With this being the case, it is not always necessary to retransmit those pixels; it may be more efficient to relocate where those pixels are located within the image. Although it is a much more complicated procedure in theory, this is essentially what motion estimation and motion compensation accomplish. They in essence predict the location of existing pixels and verify them within multiple frames, thus reducing the number of pixels that must be retransmitted.

Recording Storage Types

Most digital video recording units record the video images to a hard drive or series of hard drives before they are stored to the storage media. In other words, all of the video images are recorded onto the hard drive(s) temporarily, and then they are transferred to whatever type of storage device is being used, such as a CD, DVD, or digital linear tape. Many recorders record primarily to the hard drives and store images elsewhere only when prompted to or under certain conditions.

Recording the information in this manner gives the user several options for recording and storage. The amount of storage that can occur before user intervention depends greatly on several factors, however. Each frame requires a certain file size, which can vary greatly by the compression type and recording equipment manufacturer. The size of the hard drive is another important factor for the amount of recording that can be done, as is the number of frames per second that are being recorded.

Whichever technology a system uses, the recording options are usually the same or similar from one manufacturer to the next. As with a computer system, hard drives are utilized to initially record the images before they are sent to another storage medium. This would be like doing a tape backup on a home or business computer, but would probably occur much more frequently.

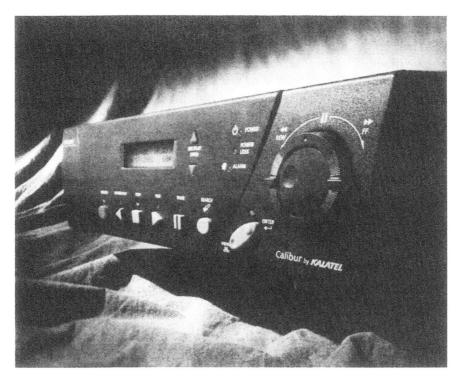

Figure 6.4 Digital recorders, such as this Kalatel DSR-2000, record images to a hard drive, allow remote Ethernet viewing, and provide an SCSI port for image archiving. (Source: Kalatel)

Hard Drive Recording

Many digital video recorders are designed to store all of the recorded video onto a hard drive or series of hard drives built into the machine (see Figure 6.4). By storing the video images directly to the hard drives, managing the file systems and images can be much less complicated than with some systems that immediately transfer the images directly to an external storage device. Uneventful video, for example, can be recorded over on a continuous loop, with the oldest video being overwritten first.

If the recorder is designed to overwrite the oldest video and record in a continuous loop, there is little need for user intervention except when there is suspicion of an incident or an alarm condition occurs. If there is an alarm or an incident that must be retained for future use, the video clip can easily be transferred to an external device so that it can be saved. These external devices can range from a standard recordable CD to an SCSI tape backup device, depending on the customer's needs.

In many cases, the programming allows several different options for external storage of video images. Most devices that record to the hard drive allow the user to manually select the image or streams of images, which must be transferred to an external device. When this technique is used, the system operator would be relied on to transfer the appropriate video clip and to verify that the proper information is completely transferred.

Digital Tape Recording

When all of the video images over a predetermined period are to be recorded to a tape, they are usually transferred in groups, as the hard drive becomes full. Relatively simple ways to achieve this are through hard drive partitioning and write definitions. With a 20G hard drive, for example, it could be divided into 20 partitions of 1G each. Every time five or ten partitions become full, it would automatically transfer the images from the hard drive to the tape. Live images would continue to record on the next group of partitions so that no activity is missed during the transfer period. In this way, the tape drive is not working all of the time, which means less strain on the unit.

The digital recording tapes can also be used for a much longer period than traditional security VHS tapes. As mentioned earlier, T-120 and T-160 tapes are rated by the manufacturer for 35 uses before replacement. DDS tapes, on the other hand, are rated by the manufacturer for 1,200 uses before replacement, something that will probably never be achieved. In addition, tapes such as DDS are electronically serial numbered. Many recorders can look at the serial number and last record date directly from the tape. If the recording is too recent, it can stop you from accidentally recording over the wrong tape.

Because the recording times are electronically stamped on each tape, it can be far easier to retrieve and review information with the digital system. With many systems you simply enter your search or review parameters, and the system will let you know which tape is required. If that tape were already loaded, the video footage from your parameters would be sent from the tape back to the hard drive for review. Because the images are digital, there is not the problem common with traditional systems. Each camera can be reviewed easily at real speed, faster, slower, forward, or reverse without the images jumping around and inserting occasional shots from a different camera. This allows the user to play an event backward and forward much more quickly and accurately than rewinding and replaying with time-lapse recorders.

This will make reviewing tapes for specific times, as compared with an access control log, for example, much more feasible and accurate. If you know that someone went through a particular door at a certain time, you could merely recall the images from the associated camera 5 minutes before and 5 minutes after that door was opened. Searching through the video archive should be much quicker, because it is based on time and not the counter position as with traditional recorders.

The rapid advancements in the computer industry have proven to be both an advantage and a disadvantage. The disadvantage has been that the advancements in hard drive size, processor speeds, and types of storage media have rapidly made systems that have been purchased obsolete. Although the older systems are usually still functional and do what the customer wanted at the time, they simply cannot do what a system created one year later is capable of.

An important consideration is to establish how long the video will be archived for future use. Because the storage media for digital systems are usually substantially smaller than VHS tapes, less physical storage space is usually required to archive the video for extended periods, such as six months or a year. The digitally archived video holds up better for these extended periods, so many organizations have even begun archiving for periods as long as three or five years. Although this may seem like overkill, for many businesses it could prove to be beneficial later if inventory audits showed missing items or in the event of liability suits. One example is slip and fall suits against hotels or casinos. Occasionally, would-be victims decide that these facilities have deep pockets and will settle quickly rather than face the public exposure. If old footage is available to show that the event is either legitimate or not, it can save the organization a lot of grief.

DV Tape Recording

Many new commercial camcorders now record into a digital format that should be mentioned here. Digital video (DV) recording format is commonly used to record digital video images to a standard Super High 8mm tape. The tape is a standard analog tape that is used in older camcorders as well, but instead of storing an analog video stream onto the tape, the DV recorders send a digital stream to the tape. The tape will record approximately the same amount of time as it would with an analog recording, but the resolution and search features are much better with the digital recording format. This recording format is not one of the four main types used within the security industry, but at least one manufacturer is using this type of file format for security-related recorders. Sony has developed DV recording with its consumer product line and has also begun tying it into the security product line.

Combination Storage Systems

Several equipment manufacturers make recorders that combine hard drive and tape backup digital recording within one unit. This allows the user several options regarding video archiving. Archiving of video can be done automatically to the tape storage media, which means that all recorded video is archived. This option is valuable for facilities that require full backup of video images for extended periods. Another option available is to archive video of alarm-triggered

Figure 6.5 Many digital recorders are actually a multiplexer and recorder in one unit. This capture unit by National Electronics is a 16-camera multiplexer with built-in digital recording, USB ports, LAN connections, alarm inputs, and relay outputs.

events automatically. If this feature is set up properly, it can be a time-saver for tape reviewing.

Many of the larger digital video systems use hard drive and tape storage, with the tape devices saving externally to the recording equipment. This is mainly an issue of physical size restrictions, because larger systems may require many hard drives and tape automation for efficiency. A 600-camera system, for example, would require constant tape changes if automation were not used. Tape machines allow the system to control, manage, and change banks of tapes, allowing for terabytes of video storage relatively easily.

Digital Multiplexing Recorders

Most of the readily available digital recording devices incorporate a multiplexer into the recording unit, although how this can be utilized and set up can vary from manufacturer to manufacturer (see Figure 6.5). Many units write the digital images to the hard drive or storage media in the same manner as traditional multiplexing units—by alternating from one camera to the next for each frame. Some new systems, however, give the user the option of setting the frame rate of each camera individually, giving the user much more control. If only one camera in a facility needs to be recorded at five frames per second and ten cameras need to be recorded at one frame per second, it would be a waste of space and unfeasible to attempt to record all cameras at the higher frame rate. By selecting each camera individually, a total record rate of only 15 frames per second is required, as opposed to an unachievable 55 frames the other way.

Digital Single-Channel Recorders

A few manufacturers have added a new type of digital recorder that may be easier to implement when upgrading analog systems. A single-channel recorder acts essentially like a digital VCR replacement. It records in the same manner as a time-lapse recorder to the end user, but instead of recording to a VHS tape, it records directly to a hard drive. With the growing size of hard drives, it allows for an increased frame rate for a longer period than traditional recorders. Digital single-channel recorders also usually allow the user to select the image quality, which affects the file size. Usually the user can choose low-, medium-, or high-resolution images, which equates to different rates of compression. A low-resolution image is compressed much more than a high-resolution image, and the file size is therefore significantly smaller. This allows for more frames in the same amount of hard drive space, extending the overall recording length.

Many single-channel recorders also allow for an external storage device, such as a tape backup system, to store important video clips. This is important for situations in which the images must be saved for evidence or future reference. In addition, many units provide a network connection for remote viewing of cameras from authorized computers.

Important Considerations

For longevity of the system, it is beneficial to choose a system that can be easily updated or upgraded to a newer storage medium. For example, if the system has the recording, multiplexing, and storage self-contained in a single unit, the entire unit must be replaced to benefit from technology advancements.

All electronic equipment is rated by the mean time between failure (MTBF). This equates to the life span of the particular components or piece of equipment and is usually measured in hours. With many of the newer components for digital video systems, the MTBF is much longer than with traditional system components, partially because the traditional equipment relies on many more moving parts than its digital counterparts. For most digital components the MTBF is typically 100,00 hours, which equates to 12 years of continuous use. Typically, a traditional recorder can be expected to last somewhere between three to five years before replacement if it is properly maintained every 10,000 hours and cleaned frequently.

Most digital products will be outdated before they no longer work, but be sure to get the exact MTBF from the manufacturer. It is important to find out the MTBF of each of the major components in the system to help determine the overall life expectancy. If the storage recording and multiplexing are all done in a self-contained unit, there are actually three major components that may fail. Be sure to find out the MTBF of the built-in storage unit or tape drive, the recording portion including the hard drive(s), and the multiplexing portion. The overall MTBF of the unit would be the same as the shortest MTBF for any individual component, not the average of all components.

Longer MTBF means less required maintenance and less frequent service replacement. If the final video output to the storage media is a digital image, this can more easily be read in other machines. Many manufacturers will make some portion proprietary to make sure that their equipment is used, but this is not always the case. A single JPEG image can be used on practically every computer, which means that an incriminating shot or an image problem can be sent to someone else to analyze—for example, as by e-mail.

Transferring the video to a digital format makes it possible to send the images over a LAN/WAN network. Many units have this option built in to allow for remote system monitoring. The biggest advantage is that a security manager can easily pull up any camera to view at any time, many times with either a Web browser or a custom graphical user interface (GUI).

If an organization has multiple facilities all on a network, the security manager can easily access the video images from any camera in any facility. It would even be possible to remotely access the images via a dial-up connection to the network or over the Internet.

One common method for transferring video from the hard drive to the storage media is to do so only on alarm. The information can continue to record to the hard drive until an alarm-triggered event occurs. The alarm trigger would then cause the information to be automatically sent to the tape drive. By using a hard drive and alarm-triggered events, the hard drive can also act as a video buffer. This allows the user to view the events just before the alarm to see what may have occurred. With alarm-triggered recording of events on traditional recorders, the recording begins immediately after an event has occurred. With the digital system, that information already exists on the hard drive, so it is simply a matter of recalling the information. Setting a buffer time in seconds or minutes or choosing a number of frames immediately before an incident to record usually does this adequately.

Writing alarm events only to the storage device means that there will be much less time required for reviewing tapes. Instead of sitting through hours of uneventful camera views, as with traditional systems, users would only be looking at those times where actual activity occurred on the screen. They would then only have to determine what was an important event and what was normal activity, a much less time-consuming task.

If the system is used with a guard force or other live user, it could be set to record the activity manually based on a user-activated trigger. This would capture events that might otherwise have been missed. An example is shoplifters in a retail environment. It is impossible for a camera to distinguish between a legitimate shopper and a person attempting to conceal merchandise. A loss prevention officer, however, could easily identify the difference and let the camera system know with a trigger, such as a pushbutton, causing the system to record the suspicious activity. The same could be done with a traditional camera system, but again the buffer period of the hard drive would provide a distinct advantage.

Digital Video as Evidence

One concern for many organizations is the use of digital video as evidence. This concern, however, is not specific to the digital systems. Whichever type of system is used, it is important to establish a chain-of-evidence policy before it is actually needed. If a criminal offense is captured in a recording, it is important to establish who will remove the storage medium from the machine, where it will be stored and maintained, and who will have access to it. Consult the corporate counsel or attorney during or before the system installation to determine what is required.

Hundreds of companies are currently making and selling digital video systems or digital video products, and there is no clear-cut standard, partially because it is such a new and growing market. Every manufacturer is trying something a little different in order to gain an advantage over other systems, and few of the products will interface with those of other manufacturers. This makes the user's equipment choice more permanent, also because of the initial cost.

CHANGING FROM ANALOG TO DIGITAL

When an organization changes from a traditional camera system to a digital one, it is not always a seamless transition. The basic operation of the newer system can be entirely different from what the security staff is used to working with, so training can be a key issue. Many organizations are under the impression that they can simply remove the old multiplexer and recorder, install the new digital unit, and then continue using the system in the same way that they always did. Although a few systems are designed specifically to achieve this goal, it is usually not the case.

If retraining the entire staff is not done adequately from the beginning, the digital system can be more of a hindrance than a help. If the security staff is reluctant to change, they may be quick to complain and find fault with the new system. It must be established initially what the added benefits of the new system are and how they affect each person who will be using it.

The security staff must also sell upper management on the benefit of changing over and should be able to show the return on investment (ROI). One way of achieving this is factoring in long-term costs of repair, replacement, upgrade, and storage media such as VHS tapes and DLT or DDS tapes. The amount of time saved by reduced reviewing time should also be factored in as cost savings.

A disadvantage that was mentioned earlier is the rapid change in technology. As new formatting, compression, and hardware are developed, capabilities will be increased. This also means that the currently installed systems may be obsolete in a short period. A look at the PC industry shows what this means: Very few people would buy a computer now with a 1G hard drive and a 60-MHz processor, but a few short years ago that would have been a hot item. The same applies to the digital video products. Most people would not choose a recorder that uses DDS2 tapes and a 2G hard drive, but a few years ago that would have been an adequate system. High-end equipment being sold today may seem outdated and inadequate in two years.

Because the technology is changing so rapidly, many new companies do not survive for long. If a company is not poised financially or with the right product for future development, it will quickly fall by the wayside. Be sure to check the company history and size, as well as customer references, to pick a company that will be around to support you.

For larger systems with 32 cameras or more, it can be quite a task to accurately review the video, regardless of which type of system is used. One of the concepts behind the creation of digital systems was to alleviate this problem, but most end users have been reluctant to set up systems that record only based on alarm activity and video motion detection triggers.

If a digital system is used to record all video, the amount of tapes can be overwhelming, or the frame rate is reduced to a point below what is usually obtainable with a traditional system. Before deciding to change to a digital system, the minimum performance requirements for the system should be established. These requirements can then be used to analyze the feasibility of the digital system as compared with traditional systems. In this way an organization can more easily determine whether a digital system is functional and cost effective compared with existing systems. In many cases, organizations have found that they are better off staying with the traditional recording techniques or just adding a few digital components and keeping the analog system in place.

A digital system can be beneficial if it is used to its fullest potential, which includes remote viewing and monitoring, alarm-activated recording, and video motion detection. If, however, an organization is still set on using the system in the same manner it always has, it will probably be disappointed.

IMPORTANCE OF ENHANCED RECORDING CAPABILITIES

The best way to take advantage of the digital system is to establish alarm triggers and video motion detection to cause certain actions within the system. In this way the recording can be set to capture only relevant footage or at least time and date stamp the footage for easier review later. Random video review will be no easier than with traditional systems in most cases, but if alarm events are marked for review, the user can go straight to those events and skip the views of an empty room.

If alarm triggers are going to be used, it is important to determine what type of occurrence will signify a notable event. For retail facilities, for example, it is not practical to use video motion detection most of the time, because the recorder would constantly be in the alarm state and recording legitimate shoppers as well as suspected shoplifters. An alternative trigger might be an alarm output from an electronic surveillance system. This would trigger if someone attempted to walk past a certain area, such as an exit, with tagged merchandise that had not been paid for.

For facilities that are subject to armed robbery, such as banks and convenience stores, a "last bill out" device in the cash drawer would be an excellent way to activate the recording without jeopardizing the safety of the employees. Silent hold-up devices would be good also if the system were set up with alarm buffers to capture the events before the alarm. For a bank, for example, it is not recommended to have a device that could jeopardize the safety of the tellers.

If the trigger could activate the buffer recording, however, the tellers could be instructed to activate the trigger only after the suspects have left or if they are clearly out of harm's way. Most bank robberies take only a matter of minutes or even seconds, so a buffer could easily be established for activation after the event that would capture everything at a high frame rate. If the digital system is constantly recording 20 or 30 frames per second and overwriting the data until a trigger occurs, the buffer would still have the required camera shots at those high frame rates.

It is important for each facility to decide what type of an event is important enough to record for each individual camera view. For example, it may not be important to record every time someone walks through the front door, but it may be extremely important to record every time someone exits through the rear door. Establishing the parameters for each camera in advance will help ensure that important events are not missed and that there is as little useless recording as possible.

The possibilities of alarm-triggered recording are only limited by the imagination of the installers and system users. Many devices could be utilized to enhance the recording capabilities of the system, from simple door contacts and motion detectors to time-based triggers from access control systems and alarms from other pieces of equipment.

CONCLUSION

The advantage to all of the advancements in digital technology has been drastically improved functionality for camera systems. In early 1999, the average digital recorder contained a 4G up to a 12G hard drive, and many recorded to DDS2 or DDS3 tapes. By the end of 1999, most digital units had a hard drive of 50G, and several contained a series of 50G hard drives.

Companies that are most concerned about what happens in their facilities after normal hours of operation are also ideal for digital systems. The system can easily be integrated with the intrusion detection and/or access control systems to monitor activities in the event of an alarm or suspicious access. Although many facilities may not have such clear-cut choices, most can be readily adapted to the digital environment with creativity and use of technology.

Much of the planning for a digital system is identical to planning for a traditional system. Camera location, field of view, lens type, and extras such as infrared illumination and housings should not change from one system type to the next. If eight cameras were needed to cover the perimeter of a building with the traditional system, they would still be required for the digital system. This may seem like simple common sense, but many users seem to believe that by going to digital they would have a super-system, capable of doing much more with much less equipment.

The digital portion of the system is merely a different type of system control and configuration. Digital systems will not increase the capabilities of the camera or a monitor. Although the digital recorder is usually capable of producing higher-resolution images, if the cameras are only capable of 330 lines of resolution, the recorded images will only look like 330 lines of resolution. Where the difference would be noticeable is if high-resolution cameras were installed but were recorded on a traditional recorder that was a much lower resolution. If the digital recorder is used with the same cameras, the images should appear much clearer with the digital system. The resolution produced will only be as high as that of the lowest-resolution device.

System design—either for replacement or for a new facility—will start the same way regardless of which system is to be used. The one thing that may cause a change in camera or pan/tilt and zoom choice is system compatibility. Not to say that some cameras are not compatible with some systems, but some features will work only if both components are from the same manufacturer. Tours and patrols, for example, are scripted actions that cause cameras to change to preset positions on a time or triggered event basis. These tours and patrols often will not operate properly if essential components are a mix from various manufacturers. Sometimes this is because of proprietary protocols by the manufacturer, and sometimes it is because one manufacturer may not even offer that possibility. If that is an important feature for the system design, it is important to verify the component compatibility in advance. In this case the manufacturers or a consultant would be the biggest help, because they have probably been asked about this issue on numerous occasions.

There are several major advantages to recording the video in a digital format. One advantage is a much higher picture resolution with no degradation. Resolution with a digital recorder is often higher than with time-lapse and real-time recorders. The resolution can be as high as that of many cameras and monitors, meaning that the recording equipment is no longer the restrictive factor for excellent picture quality.

With digital recording, reviewing previously recorded video is much more efficient. With the traditional multiplexed video, it is not really possible to fast forward or rewind a particular scene while viewing. With digital recording, the video can be set to start playing at a predetermined time, and fast forward or rewind can be used to search for a particular person or incident. Because most digital systems are computer based, video footage can be navigated much more easily than in traditional systems.

An additional benefit of digital storage is the space requirement for archived tapes. To store 60 days of videocassettes would take quite a bit of room, but 60 days of digital tapes would easily fit into a desk drawer.

$$7$$

Enhancing Recording Capabilities

REVIEWING TECHNIQUES

Perhaps the most common application for modern camera systems is to record the video from the cameras for future review or evidence. Many organizations have systems set up to record all of the camera video but no one to view the cameras live. Many convenience stores, for example, have recording systems, but the staff is too small to have someone observing live activity. In this environment, the only live view is often from a monitor at the register that allows customers to see the camera view.

Security camera systems are usually set up to record all of the cameras all of the time. If there is 100 percent recording, that equals 8,736 hours of recorded video each year. If someone is tasked to review this video footage and review takes 10 percent of the record time, that means 873 hours per year of tape review. That adds a significant annual cost to the security budget to have the staff available for tape review. Based on a 2,080-hour work-year, tape review would tie up one employee for nearly half of a year.

Because of the labor involved, most organizations do not perform 100 percent tape review, which can be just as much of a concern. Most organizations only review archived video when a particular incident is brought to the attention of the security staff. Although major incidents, such as a break-in, would be noticed, subtler incidents could go undetected and never be reviewed on the videotape. An accumulation of these smaller incidents could amount to a substantial amount of annual lost revenue for the company. In a retail environment, for example, items missing from the inventory could go undetected for some time or be written off as bookkeeping errors, when in fact they have been stolen. Cash register shortages, invalid returns, and other scams could be considered just mistakes when in fact they could be organized theft.

Whether performing total review or incident-based review, the organization faces a potential loss that could be avoided. A loss through unnecessary

Figure 7.1 Some areas can be difficult to monitor effectively by simply recording 24 hours a day. Triggering devices can often be used in such areas to help reduce the amount of irrelevant video footage recorded.

labor or undetected theft will ultimately impact the bottom line of the company as well as the effectiveness of the security department.

For periodic, incident-related, and total review, the effectiveness always stems back to the fact that 100 percent recording means far too much video to review properly. The inherent flaw is that continuous recording produces hours of footage with no activity whatsoever. Whether using traditional videotapes or digital recording, this space and time could be better utilized if important activity could be predicted (see Figure 7.1).

Although it is impossible to predict exactly when suspicious activity may occur, it is possible to note the exact moment that certain types of activities occur. Because it is possible to monitor and track activities such as movement and doors opening, it is possible to reduce the amount of time that the cameras are recorded. Any reduction like this will also decrease the time required to review tapes and make it much more simple to locate a particular incident on a tape.

WHAT ALARM TRIGGERS DO

One way to enhance the recording capabilities of the camera system is with the addition of alarm detection equipment and *alarm triggers*. Most multiplexers, recorders, and matrix systems have alarm input capabilities, as shown in Figure 7.2. Each of these devices has a place to connect alarm-triggering equipment, and each uses those inputs to affect the recording and/or display properties of the camera system. They also usually have relay outputs, which can be triggered to enhance the recording capabilities as well.

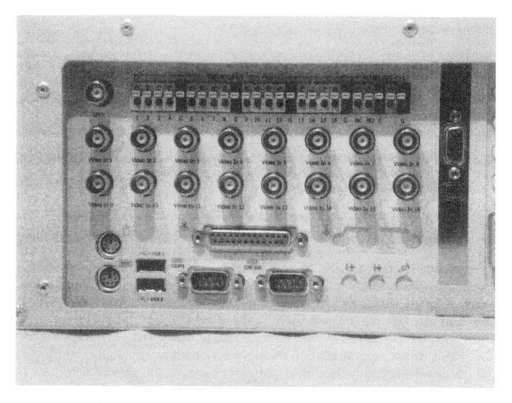

Figure 7.2   Recorders, such as this digital unit by Synectics, often have alarm input connections on the rear panel to help enhance the recording capabilities of the system. This unit has 16 alarm inputs and one alarm output.

Figure 7.3 Door contacts can be used as an alarm trigger to indicate that the door has been opened. If there is no activity at the door, then continuous recording would only produce hours of unnecessary video footage.

An alarm trigger would be any device, such as a door contact or motion detector, that lets the user or a piece of equipment know when a certain type of event occurs. A door contact informs the user or alarm panel when a door is opened or closed. A motion detector signals when there is movement in the area. A duress button signals that someone requires assistance or is in trouble. What is done with the information provided by these devices can greatly improve the effectiveness of the camera system if they are used properly.

Many devices can be used to enhance the recording capabilities of the system. The most frequently used items are door contacts and motion detectors. In retail and banking environments, another common trigger is a holdup or duress button. For exterior cameras or cameras in large indoor areas, microwave and photobeam detectors can be used, as well as fence protection systems, seismic detectors, and vehicle loop detectors. Virtually any alarm device used on an

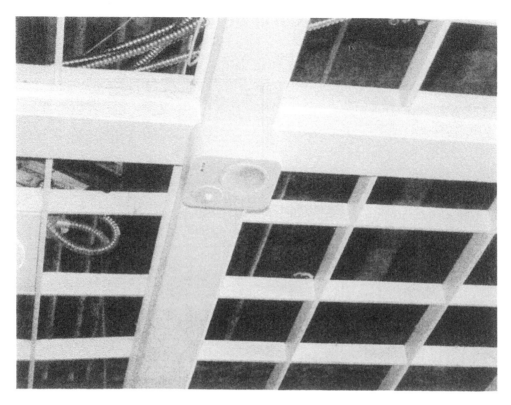

Figure 7.4 Motion detectors are an excellent device to trigger enhanced recording only when there is activity in the covered area. This unit by C&K is a ceiling-mounted 360-degree detector that covers a radius of 25 feet in any direction.

alarm system can be used as an alarm-triggering device with a CCTV system. Figure 7.3 shows a typical door contact installed, and Figure 7.4 shows a 360-degree passive infrared and microwave motion detector.

CHANGING RECORDING SPEED

When the alarm devices are connected to a multiplexer or a video recorder, the receiving device must be told what to do when an alarm occurs. The most obvious change that can be made when there is an alarm is to increase the recording speed. This means that when an alarm occurs, the recorder will record more frames per second onto the tape than when there is no alarm. The advantage to

Configuring Alarm Devices

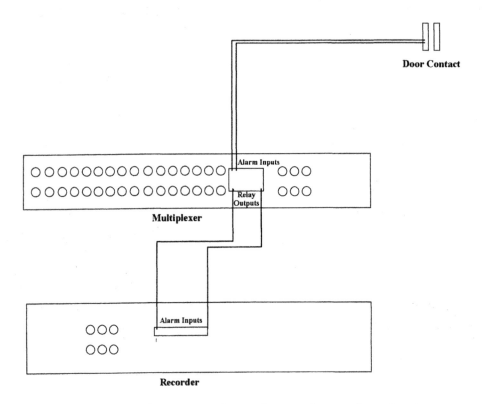

Figure 7.5 This diagram shows one method of changing the record speed and camera record sequence with a single alarm trigger.

this option would be that the recorder can be set for a relatively slow record rate for normal operation, and, when an alarm is detected, the frames per second on the camera can be increased to show more detail of the activity. This means that a tape will contain fewer frames of no activity or that the tape will record for a much longer period. When the alarm resets, the recorder would return to the regular recording rate. Every time an alarm is received, the recording rate would change to the quicker rate and would change back when all of the alarms are reset (see Figure 7.5).

One way to trigger the change in the record rate is that an alarm trigger would need to be connected to one of the alarm inputs on the rear panel of the recorder. If the recorder is being used with a multiplexer, this technique will increase the recording rate of all cameras being sent to the recorder. For a low number of cameras or if the cameras are all covering the same general area, this may be adequate; however, it is usually desirable to increase the record rate on one or two cameras, not all of them.

To change the record rate on specific cameras requires more planning than simply connecting a triggering device to the rear of the recorder. Alarm inputs to the multiplexer, relay outputs from the multiplexer, and alarm inputs on the recorder will all need to be utilized. To increase the record rate for specific cameras, two separate functions must occur: (1) the record sequence of the multiplexer must be changed to increase the frames from the camera, and (2) the recorder must be triggered to change the recording speed. In order to achieve both functions with a single alarm device, the alarm cabling must be connected in a particular manner. The alarm device must first be connected to one of the alarm inputs on the multiplexer. Then the multiplexer must be programmed to associate this alarm with a particular camera or group of cameras. To trigger the recorder, a pair of wires must be connected from a relay output on the multiplexer to the alarm input of the recorder. By configuring the system in this way, the multiplexer can change the sequence of the frames being sent to the multiplexer, as well as change the number of frames per second. The recorder can then also change the number of frames per second so that it matches the output of the multiplexer. This will ensure that there are adequate images from cameras that have an alarm, while also reducing the frames when there is no activity.

CHANGING FRAME FREQUENCY

To better understand the reasons for this type of configuration, the reader should understand the possible changes that can be made with the multiplexer when an alarm event occurs. Utilizing the multiplexer alarm inputs can have the biggest effect on what information is recorded and what is ignored. Although the options may vary from manufacturer to manufacturer, most offer alarm inputs and video motion detection as a means of enhancing the recording capabilities of the system.

Exclusive Recording

One of the alarm interface options available with most multiplexer manufacturers is the ability to change the recording to *exclusive recording* when an alarm

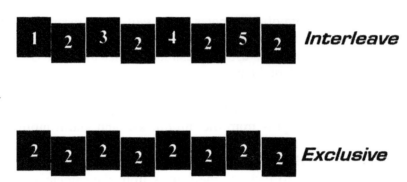

Figure 7.6 Many multiplexers offer exclusive and interleave recording to enhance the recording capabilities of the system.

event occurs. Exclusive recording means that only those images from the camera or cameras associated with the alarm event are sent to the recorder. This means an increased recording rate for those cameras and no recording of cameras without an alarm condition. For example, with no alarm conditions, if a system were recording ten cameras at a rate of five frames per second, each camera would be recorded at a rate of one frame every 2 seconds. If an alarm event occurred that was associated with only one camera, it would then be recorded at a rate of five frames per second, and the other nine cameras would not be recorded at all. This type of recording is most beneficial when the cameras are set to view general activity and the alarm events signify the most crucial recording requirements in the facility (see Figure 7.6).

Interleave Recording

Many multiplexer manufacturers offer an alternative to exclusive recording that can often be more beneficial. With *interleave recording,* the camera associated with an alarm event is recorded every other frame. That means that when there is an alarm event, the camera or cameras associated with that alarm event are recorded more frequently, but the other cameras are still recorded as well. For example, with no alarm conditions, if a system were recording ten cameras at a rate of five frames per second, each camera would have a record rate of one frame every 2 seconds. The standard recording sequence would be 1, 2, 3, 4, 5, 6, 7, 8, 9, 10, and then would repeat. If an alarm event were to occur that was associated with camera 2, every other recorded image would be of camera 2. The new record sequence would then be 1, 2, 3, 2, 4, 2, 5, 2, 6, 2, 7, 2, 8, 2, 9, 2, 10, 2, and then would repeat. This means that one camera with an associated alarm event would be recorded five times in every 2 seconds and nonalarmed cameras would be recorded only one time in every 4 seconds. In many situations this can be much more beneficial than the exclusive recording, because each camera is still being recorded.

Most manufacturers that offer interleave recording allow the user to choose between exclusive and interleave. It is possible to use both options on the same system if they are used in conjunction with scheduling. In other words, it may be necessary to use interleave recording during normal working hours for many businesses and better to use exclusive recording in the evenings and on weekends. This choice is all a matter of programming the multiplexer properly and ensuring that the multiplexer program coincides with the recorder program.

ALARM INPUTS

Many camera locations are ideal for the use of alarm triggers. These are areas with either specific types of activities that occur or with a low number of occurrences. An emergency exit door, for example, is a great place to use a door contact for an alarm trigger. The traffic through the door is minimal, and the security staff will want to know any time an emergency door is used. If alarm triggers are not used with this type of application, the videotape will contain hours of video looking at the door with absolutely no change. That means that this camera location would be more difficult to review and also that it is using valuable frames per second that could be better utilized by a camera with more activity.

Looking at the emergency door example, if a door contact were installed on the door and connected to the alarm input of the multiplexer, that camera could be removed from the regular recording sequence. That means that no images from that camera would be recorded unless someone physically opened that door. The camera could also remain in the regular record sequence but be set to record more frequently when the door is opened. If every camera in the record sequence were similar to this situation, the frame rate and record speed could be reduced so that more elapsed time would fit onto a single tape.

One drawback of using a door contact as the primary alarm trigger is that once the door closes, the alarm is reset. When the alarm resets, the recorder and multiplexer will return to their normal recording mode. Although it may be effective to show the actual person walking through the doorway, the system may return to the normal record mode too quickly and may not show what occurs immediately after. In many situations it may be important to have enhanced recording any time that a room or area is occupied. If that were the case, a simple door contact would be inadequate as a trigger. A timer module could be added between the door contact and the multiplexer alarm input as one option, but this would require additional equipment and proper configuration. If someone were to open a door and immediately leave, extending the alarm record time with a timer module would cause unnecessary recording and again waste space on the tape.

Two other methods of event triggering can be used that help ensure that only useful video is obtained. First, instead of using just door contacts as the alarm trigger, a motion detector or group of detectors could be used to trigger enhanced recording any time there is activity in the area of concern. With the motion detectors, any time there is movement in the area, an alarm would be generated that would cause the multiplexer and recorder to go into the enhanced recording mode. The multiplexer could be set to stay in the alarm mode for a period after the motion detector resets as well. This additional time should be set long enough that if someone is consistently moving in the area, the motion will alarm again before the additional time expires. This will prevent the multiplexer and recorder from continuously toggling between the regular recording and enhanced recording modes. The time should be short enough, however, to prevent a lot of unnecessary recording when no one is in the area.

If multiple motion detectors are used to trigger cameras, it may be possible to track someone through the facility, particularly after business hours. Figure 7.7 shows a typical setup. The motion detector for a camera in one area would trigger to start the enhanced recording. If that person moved to a different area of the building, the next motion detector could trigger enhanced recording on a different camera covering that area. With the proper placement of cameras and motion detectors, it would be possible to map the entire facility and accurately follow an intruder at a higher record rate. By utilizing this enhanced recording, the standard recording could be set to a slow record rate, meaning a single tape could last for days. Reviewing would be much simpler, because there is less tape to review, and the reviewer could simply look for the alarm triggers.

VIDEO MOTION DETECTION

While alarm triggers are an excellent way to enhance the recording capabilities of the camera system, they do require more equipment, cabling labor, and programming. Many multiplexer manufacturers include an additional feature that

Figure 7.7 Pictured are two cameras, each with a motion detector for alarm event triggering. These cameras are located at an intersection of two building corridors. Any motion in either corridor will trigger enhanced recording for the associated camera, making it possible to track activity throughout the facility.

can help with enhanced recording that does not require any additional equipment.

Video motion detection is the ability of the multiplexer to analyze the image from a camera and determine when there is movement within that image. This is accomplished by noting changes in the pixels of the image, triggering an alarm when there are enough pixel changes to indicate motion. The reliability of the video motion detection depends on each unit and whether masking and sensitivity adjustments are possible. Video motion detection units that can be incorporated into any camera system are also available separately.

A big advantage of video motion detection is the ability to enhance the recording capabilities any time movement occurs in a camera viewing area (see Figure 7.8). If alarm-triggering devices alone are used, the recording may display

Figure 7.8 This large parking garage utilizes nearly two dozen cameras to monitor activity in any area throughout the garage. During peak hours, recording is continuous on all cameras. With video motion detection enabled on the multiplexer, however, corridors with no motion record less frequently. The store can more efficiently monitor traffic flow, incidents of concern, and vehicle tracking.

the instant the door was opened but miss the illegal entry technique used to gain entry or the suspicious activity that occurred once the door was closed.

One important requirement when using video motion detection is the ability to disregard or mask areas of the display that can cause false triggers. For example, when using this feature with outdoor cameras, a tree in the field of view would probably trigger enhanced recording every time the slightest breeze blew. Because the video motion would constantly be activated, this would completely defeat the purpose of using it. To eliminate this problem, it should be possible to block the motion detection in certain areas of the image. Most multiplexers provide a video motion setup that allows the user to go through the image display and turn sections on and off. The entire area over the tree could be disregarded so that only activity in the desired viewing area causes a change to the recording.

When setting up the video motion detection zones or areas, the user should pay close attention to which areas of the image may have activity that should be recorded. Areas such as ceilings and lights usually should not be included in the motion area. Fluorescent lights in particular should be masked, because they can easily cause false motion triggers. Areas such as walls may also be masked if there is no possibility of something occurring in front of them. The video motion should only be used on the areas where enhanced recording is required. Users will often activate the video motion feature without properly programming the detection area. Undoubtedly, this will lead to false motion triggers and complaints that the video motion feature is unreliable or too sensitive. Properly programming each camera view will eliminate the majority of the false triggers and make the enhanced recording much more reliable.

Once the detection zones are established, it is important to set up the sensitivity for each camera. Most manufacturers allow for multiple sensitivity levels with the video motion detection to further eliminate false triggers. The number of sensitivity levels and their description will vary from one manufacturer to another and may be an important factor when choosing a multiplexer. Some multiplexers will have as few as three sensitivity levels: high, medium, and low. Others may have as many as 99 levels, each given a number designation of 1 through 99. Still other manufacturers use a sensitivity setup that includes indoor high, medium, and low, and outdoor high, medium, and low. Figure 7.9 shows a typical outdoor application.

Providing a high number of sensitivity levels means that the user can set up the system to have the highest sensitivity possible without having numerous false activity triggers. Setting up each camera view for the best accuracy may be time consuming initially but will drastically save time in the long run with reduced tape reviewing time. If a multiplexer has a low number of sensitivity levels, one level may have frequent false triggers, whereas the next level could miss important activity. This may be sufficient if the video motion detection is used in conjunction with alarm-triggering devices for enhanced recording, but in many cases it simply leads to user dissatisfaction with the system.

Directional Video Motion Detection

Although most multiplexers are similar in their video motion detection capabilities, one manufacturer has added a new feature that can be helpful with enhanced recording capabilities. I have attempted to refrain from using names of

Figure 7.9 This outdoor roof camera is installed to monitor activity on the roadway and parking area outside the facility. To use video motion detection effectively to enhance the recording capabilities, only those areas of concern should be monitored for motion. Other areas, such as the trees and the roadway off the property, could be masked to reduce the number of triggering incidents.

manufacturers and specific pieces of equipment, but this feature is an important development in the security industry that should be addressed.

ATV has begun to include a new feature with many of its multiplexers called *directional video motion detection*. Directional video motion detection has all of the same characteristics and programming options as standard video motion detection, but it also provides the user with another triggering option. It allows the user to specify which direction the motion must be traveling in order to activate an alarm trigger. For example, the system could be set up so that video motion would only become an actual alarm if someone or something moved across the screen from left to right. This added feature provides countless new possibilities and drastically increased accuracy when looking for certain types of events.

The directional motion detection also has several programming options, which further enhance the recording capabilities of the system. First, virtually any direction can be specified as a trigger, such as left to right, right to left, motion up, motion down, or even motion diagonally within the image. Second, the direction of motion can be synchronized with a schedule as opposed to always set for the same direction. For example, in a school, traffic flow in the morning is generally into the building, whereas in the afternoon traffic flow is generally out of the building. The system can be programmed so that in the morning any motion that shows someone exiting the building, regardless of how many people are entering, will cause an alarm. Through the same cameras in the afternoon, the system can be set to trigger on any movement that shows someone entering the building. Normal flow of students would be ignored, and no alarm trigger would be generated, but the moment that anyone goes against the flow of normal traffic, the recording is enhanced to provide more images from the associated camera.

Just using this type of example, the applications are endless. Traffic flow in parking lots, parking garages, one-way streets, restricted-access areas, and much more could easily be monitored. Museums and places with high-value items could set the system to ignore all normal flow of people but trigger instantly if an item is picked up or moved in a certain direction. Shipping and receiving areas could ensure that items leaving exit only through shipping and items entering come in only through receiving. The use of this feature is limited only by the creativity of the user.

ENHANCING RECORDING WITH PTZ CAMERAS

To this point, this chapter has covered using alarm triggers and video motion detection to enhance the recording capabilities of the system. The techniques described thus far refer mostly to fixed cameras and are not effective if a camera has pan/tilt and zoom (PTZ) capabilities. One of the problems with the PTZ camera is that the user cannot always be assured that it is focused in the right direction or on the proper viewing subject.

Many PTZ cameras, particularly the newer dome-type cameras, have a feature that can eliminate this problem and make sure that the camera is always aimed in the direction where it is needed. Most of these cameras have the capability of setting up what is known as camera *presets*. A preset for a PTZ camera is

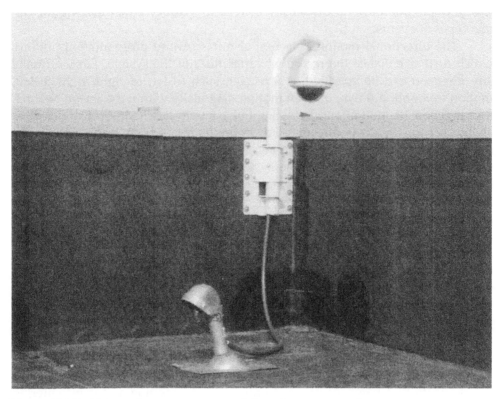

Figure 7.10   This outdoor camera is a pan/tilt and zoom unit with up to 32 presets and alarm inputs. Mounted at the corner of the building, it can easily be used to change the field of view for up to 32 predetermined areas when an alarm occurs in that area.

a predetermined direction and/or viewing area for a camera, which it can return to at any time. One of these presets is often known as the camera park preset. The park preset is the position that the camera will return to after a certain period in which the camera is not moved or utilized. This will ensure that the camera is not left looking at a useless viewing area and forgotten by the system user. Figure 7.10 shows a typical outdoor PTZ camera installation.

Presets can be used effectively with alarm-triggering devices to track activity in a facility or within the range of a camera. Many cameras have the potential of up to 64 presets per camera, and many even have the alarm inputs right at the camera housing. If utilized properly, the PTZ camera can become as effective as many fixed cameras covering the same area. If used improperly, however, the PTZ camera can essentially become useless and could remain in motion almost continuously.

Utilizing Presets

Presets for PTZ cameras are an effective way to gain the most from the camera with the least amount of human intervention. Once each camera is programmed and the terms of the presets are programmed with the control equipment, the user will not need to do much more with the camera. Although users can still manually manipulate the camera position if necessary, all of the camera functions can be automated to react in certain ways under certain conditions.

To properly utilize the presets of a PTZ camera, the user must first know what function and multiple viewing areas will be associated with that location. An outdoor camera, for example, could be used for several different functions and may require that the camera is moved and/or zoomed to view the various scenes desired. Perhaps the camera is used most often to observe traffic patterns in a parking lot. On occasion it may be necessary to look at an entry door or emergency exit to see what is going on there. Still other times it may be necessary to take a close look at an area immediately around an emergency call station if someone has requested assistance. For an operator or user to change this camera view frequently would require time and take away from the rest of the camera viewing as well as other job duties that might be required. If, however, the multiple viewing possibilities were automated and set to move with predetermined triggers, the effectiveness of the user would increase drastically. If an alarm were to come in and change the camera view to one of the presets, the alarm trigger could also cause that camera view to be displayed automatically on a spot monitor of the multiplexer. A relay output of the multiplexer could also be activated to sound an audible alarm, drawing the user's attention away from other tasks to the camera view.

Once the viewing possibilities have been established, the next step is to program the actual presets for each camera. Each manufacturer provides detailed information about how to program the presets, and most are similar. From the programming mode, the user will enter preset programming, move the camera and lens to the desired location, and follow the manufacturer's directions to lock it in. Some manufacturers also provide the ability to insert a camera title for each preset to be displayed on the viewing screen. This will make sure that users are aware of exactly what they are looking at.

Once the presets are established, the alarm triggers must be determined (see Figure 7.11). Door contacts, motion detectors, photobeams, and duress buttons can all be used as triggers to initiate the camera movement. Programming in response to these triggers is the section that will tell the camera exactly what to do. Depending on the equipment, this programming could be part of the multiplexer, matrix switcher, PTZ controller, or even the camera itself.

Figure 7.11 This emergency call box by Louroe provides an alarm trigger at the receiving end, which can easily be tied to the camera system. A PTZ camera preset would then be associated with the alarm trigger to cause the camera to focus in on that area any time the emergency button is pressed.

In addition to having the camera move under various alarm conditions, some manufacturers provide the capability of having the cameras perform tours or patrols. This means that the system can have the camera(s) go through a string of preset positions in a particular sequence. Each camera can stay in a preset position for a predetermined period, and then change to the next preset position. After the camera has finished the string of presets, the next camera can pick up and go through a string of presets. Using this technique, it is possible to sequence through a series of camera views and a series of cameras, completing a tour of the entire facility.

The name *tour* or *patrol* stems from the fact that it was designed to simulate a tour or patrol of a human guard without requiring a person to walk the rounds. Because the entire procedure can be automated, it can be more effective than an actual person performing a tour and can be a valuable tool to assist a guard force. A camera system cannot intercede on an incident that may be occurring, however, so the camera tour cannot effectively replace a guard. A human would still need to respond to any incident that required immediate attention.

CONCLUSION

There are many ways to enhance the recording capabilities of any camera system. Unfortunately, many of these techniques are often overlooked or deemed as not worth the effort that is required to implement them. When no enhancement techniques are used, the organization usually ends up with either hours of useless video that is never reviewed or no recorded video at all. If some or all of these techniques are used to their fullest potential, system users are usually much more satisfied with overall system performance, and the system actually becomes much easier to use.

Most of these techniques can be effectively incorporated into existing camera systems with little additional equipment and labor costs. It is hoped that readers will see the benefit of many of these techniques and put them to use. If done properly, they can show an immediate benefit to the organization, the security manager(s), and the system users, ultimately affecting the security budget and bottom line with increased system confidence and reliability.

Covert and Overt Cameras

For quite some time, covert or hidden cameras have been used successfully by law enforcement and private security to capture illegal activity. Unfortunately, the covert cameras themselves have also been used to perform illegal activities (see Figure 8.1).

Hidden camera usage is one of the most controversial topics among security professionals and private citizens alike. Installing covert cameras can subject the user to a legal minefield that can be costly at best even with proper legal guidance.

This chapter examines the variety of hidden cameras and a few of the legal pitfalls of using them. This chapter does not nearly cover all of the legal aspects of using covert cameras, nor could it. Each case in which a covert camera could be used must be independently evaluated, and even then could be a problem legally.

USING HIDDEN CAMERAS

There are many characteristics of cameras that help divide them into various categories. Monochrome and color cameras, for example, have a distinct difference and distinct applications. Likewise, covert or hidden cameras have distinct applications and purposes compared with their "plain view" counterparts.

Hidden cameras are often overused, and their applications are often over-glamorized. Although many situations may require covert cameras, they are often used in situations that could cause more trouble than help. Covert security cameras have begun popping up worldwide in nonsecurity and often illegal activities. It is not my intent to cover all covert camera applications, but there are several applications where covert cameras should and should not be used in the security industry that should be addressed.

Figure 8.1 Pictured are a few covert and overt cameras from Kalatel, along with two monitors.

Covert cameras by design should be used solely as a means of gathering evidence. Because they are designed to be undetectable by the persons being viewed, there is essentially no deterrent factor with their use. The exception to this rule would be a facility that uses only covert cameras, and employees know that cameras are being used but do not know specific locations. Some facilities use cameras in this manner, but the impact should be carefully considered before deciding on this approach. Employees assuming that cameras are covering certain areas can be both a help and a hindrance. Because they are unsure of exact locations, they may be less likely to steal or do anything else illegal. Most likely, this will be a short-lived benefit, however, because eventually most employees will forget about the cameras. Because they do not see the cameras as a daily reminder of security measures, the cameras eventually become less of a concern and eventually are not even thought about.

Another important factor when using all covert cameras is the liability issue from two different standpoints. First, if employees think that there are hidden cameras throughout the facility, they may have a false sense of security in areas without video coverage. Because they are not sure of specific locations, they could assume an area is protected, when in fact it is not. If an incident were to occur in such an area, the facility could possibly be opening itself up to potentially large lawsuits. In addition, by not informing employees of coverage areas, the employees could easily raise concerns about invasion of privacy. It may be a public or private facility with coverage patterns similar to a traditional system, but the mere fact that the cameras are not visible could lead employees to believe that they have a higher level of privacy than they actually do. The company may be able to eventually show that an employee should have no reasonable expectation of privacy within the coverage areas, but the cost of litigation and/or settlement could still be expensive. Any company considering covert cameras should consider these possibilities before making a decision and consult with legal counsel.

Probably the best practical application of covert cameras is to enhance an active investigation. Whether it is an internal company investigation or a private investigation, video evidence is a great way to eliminate all doubt. When used in an investigation, cameras should be used properly and not as a fishing expedition. In other words, installing a covert camera should be the last step, not the first.

A practical example is consistent cash register shortage in a retail environment such as a convenience store. Often, more than one clerk will have access to a register during the working day, so it may not be clear who is shorting the register. Review of when the shortages occur and which employees with access were working may show that the loss is consistent with a single employee or at least a limited few. At this point, a covert camera may be helpful to have evidence of the loss, how it is occurring, and who is responsible. If the camera were installed first, the investigator would be required to review hours of videotape to find the loss. If it is installed after narrowing down the suspects, the investigator can focus on those times when the suspects are in view and more easily identify the cause.

Enhancing with Hidden Cameras

Covert cameras can also be effective as an enhancement to a video system. If most cameras are overt or in plain view, a hidden camera could be helpful to view high-value items or areas, or higher-security areas. Because the primary cameras are out in the open, there would not be an expectation of coverage unless a visible camera is in the area. This application often applies to such areas as retail cash offices, safe rooms, and vaults. These are areas where there is clearly no expectation of privacy by employees, a potential loss could involve a substantial amount of money, and everyone who uses the area is fully aware that it is a high-security area with the highest level of protection.

For this type of application, the employees who work in the safe room or cash office already are aware that they are always under close scrutiny. The exact camera location or even the fact that one is installed may or may not be disclosed to the employees. This is a matter that would have to be discussed with the company attorneys and human resources.

Figure 8.2 illustrates a camera inside a motion detector.

HIDDEN CAMERAS AND PRIVATE OFFICES

Any time hidden cameras are used, privacy laws, civil rights, and employee rights are usually called into play. Hidden cameras have been a sensitive issue in the history of their use, and the acceptance of their use depends on many factors. Camera locations, type of facility, type of recording, and camera purpose can all weigh heavily on whether a particular camera is acceptable.

National Labor Relations Board

In 1997, the National Labor Relations Board (NLRB) ruled that employers must bargain with unions over the installation and placement of hidden cameras within the employer's workplace. This decision resulted from an intervention between the Colgate-Palmolive Company and its labor union. The NLRB ruled that the installation and use of hidden cameras is "germane to the work environment and outside of the scope of managerial decisions lying at the core of entrepreneurial control," so it requires that an employer must bargain over them.

The entire debate on cameras and their use seems to hinge on the phrase "reasonable expectation of privacy." If a person is in an area where a reasonable expectation of privacy exists, then it is reasonable that the person should not have to be under the watchful eye of a camera system. Bathrooms and changing rooms, for example, afford a person a complete expectation of privacy. Any use of a camera in these areas would not only be unacceptable and inappropriate,

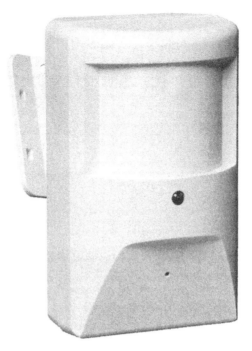

Figure 8.2 Pictured is a camera inside a working motion detector.

but it would also be a liability to the company and/or responsible party. Unfortunately, many applications are not as clear-cut as this one.

Companies in the private sector are often the most criticized for their use of covert cameras, whether that use is legitimate or not. To understand the controversy in the situation, it is important to look at the situation from both sides of the fence. From the employees perspective, they are assigned a computer, desk, and perhaps a "private" office. Many employees believe that what they keep in their desks or on their computers and what they do in their offices is their business and no one else's. Many civil liberty groups may agree with the employees on this view, depending on the situation. Many people believe that an employee in an office behind closed doors has a reasonable expectation of privacy and therefore should not be subject to video surveillance. The same argument holds true for other areas and may even be addressed in writing by groups such as unions. Warehouse areas, for example, may be protected by union guidelines for general purposes. This may not hold true if the video is part of an investigation on a particular individual.

Figure 8.3   Pictured is one type of covert smoke detector camera similar to what may have been used and cited in *California* v. *Drennan*.

On the other side of the issue is the right of the employer. It is the corporate contention that while the desk, computer, and office have been assigned to the employee, they remain the property of the employer. The employee receives compensation in exchange for his or her work time within the facility, and that time then also belongs to the employer. The company would argue then that the employee should have no reasonable expectation of privacy because the time and materials all belong to the employer. Most employers would not press the issue if an employee takes care of private matters on company time occasionally. If, however, the employee is consistently using four hours per day for non-work-related tasks, this could be conceived as theft (of company time). Because that employee was compensated for performing work and he or she did not actually work, the employee in essence stole four hours of compensation. Why is this relevant to the camera system? A camera in the company-owned office is then being used to protect the company assets from theft, whether of time or materials.

As if this issue weren't confusing enough, some situations could make this debate more complex. For example, many of the applicable laws vary from state to state. What has been deemed as acceptable in one state may be illegal in another.

In California, in 1999, for example, was a case that at the time seemed to affect the perception of "reasonable expectation of privacy" (Levine, 2001). In this case, a hidden camera with a recorder was placed in a school principal's office. The school superintendent had the camera system installed professionally by a Redding, California, alarm company, and each day the maintenance chief would bring the previous day's tape to the superintendent. The system was installed under the guidelines of the school district's attorney and with the knowledge of the school board director.

In this case, the police responded to an anonymous tip and discovered a camera in a fake smoke detector attached to a recorder above the boys' restroom ceiling (see Figure 8.3). The superintendent was charged with felony eavesdropping. It took the jury less than 1 hour to find the defendant guilty, and the court later sentenced the superintendent to three years of felony probation with ten days served in the county jail and fines and restitution totaling $7,010. The ten days in jail was stayed pending the outcome of an appeal.

When the case went before the Court of Appeals of the State of California, the decision was overturned (*California* v. *Craig Boyd Drennan*, 2000). Although overturning the conviction means that no precedent was set on videotaping in private offices, a couple of comments in the appeals court document leave some room for future cases as well as help establish guidelines for the security practitioner. For example, one of the reasons given for overturning the felony eavesdropping conviction was the method of recording that was used. In California, the penal code refers to the recording of "confidential communication." The jury took this to mean any and all communications whether verbal or nonverbal. The appeals court noted that the videotape recorded only one image every 3 seconds, which was seen as a series of still photos instead of a videotape. Because the images were so far apart in time, it was noted that it is doubtful that any communication could be deciphered, such as by lip reading, and therefore was not eavesdropping. It was also noted from the camera viewing angle that the intent was not to monitor the principal's activity but to watch the files that had been rifled through in the past. It was the superintendent's testimony that at no time did he observe a two-party communication on the videotape.

This overturned conviction opens up a few questions for future cases. For example, if a recorder had been used that recorded 20 frames per second and it had been possible for someone to read the lips of any parties who may appear on screen, would that be enough to result in a felony conviction? What about five frames per second?

Evidence presented at the original trial showed that the office was used on many occasions for private communications and activities to include investigations by the local police department. This distinction also makes the case slightly different from the typical employee's office or work area. In California, this case established that the principal has a reasonable expectation of privacy within his office. It is possible that if the superintendent had consulted with the principal before hand, the camera could have been a valuable tool in the investigation.

What this means to the responsible members of the security department is that they must be accountable for their actions. Those parties who are responsible for the design and installation of a camera system must be 100 percent sure that they do so well within the law. Because there are so many gray areas, it is

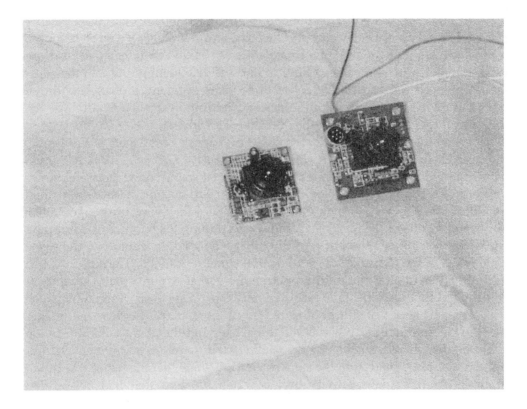

Figure 8.4 Pictured are two board cameras that are typically used in covert installations. On the right is a camera with a built-in microphone for audio monitoring or recording. This type of camera is no longer readily available in the United States largely because of the legalities of audio recording.

important to consult with legal counsel and security consultants who are familiar with any federal, state, and local laws. Remember that if it sounds questionable to a security professional, it will probably not hold up in court.

COVERT CAMERAS WITH AUDIO

Probably the most controversial cameras are those that are covert and have audio capability (see Figure 8.4). This is such a big issue that the FBI in conjunction with the Security Industry Association has begun to crack down on the manufacture and sale of these devices. They are still readily available, however, from overseas manufacturers and via the Internet.

The use of audio with a camera system can be a problem whether the camera is hidden or not. Voice recording laws vary completely from state to state and should not be taken lightly. In the state of Maryland, for example, it is law that all parties being recorded must know they are being recorded in advance. This law was brought into play against Linda Tripp for her famous conversation recordings with Monica Lewinsky. (It is my understanding that by merely including these two names in the book, sales should be in the millions!)

Privacy with conversation is much more critical than it is with video. Adding audio to the camera system can increase the chance of lawsuits from employees and customers. With conversation, regardless of where that conversation is held, there is a much clearer reasonable expectation of privacy. For example, take any two people sitting in a football stadium watching a game. If a television or security camera were to zoom in on those two people and either record to security tape or broadcast it on television, it is not likely that it could be seen as an invasion of their privacy. If, however, the two people are engaged in a private conversation, they have every reason to believe that their conversation remains private. If the television camera zooms in and records their conversation as well as their picture without their consent, they would have every reason to be angry and possibly even file a suit against those responsible. In an office environment, a private conversation among two or more people is private regardless of where it takes place.

If audio is recorded in a facility, usually it must be visibly posted that those who enter are subject to recording. If audio recording is necessary, it should be planned carefully with legal counsel, law enforcement, and anyone who may have a say in its use. At no time should a security manager take it upon him or herself to use covert audio and video recording in the course of an investigation. If an investigation is crucial enough that audio recording is required, calling in law enforcement may be in order. Audio recording should only be done by authorized people and in accordance with all federal, state, and local laws.

The most important thing to remember when considering covert cameras is to look at all possible scenarios and aspects of the situation. Be sure to consider all potential liabilities and risks and discuss them extensively with legal counsel. Although camera systems can be beneficial, improper use of covert cameras can be career ending and could possibly lead to jail time. Ignorance of the law is no excuse, and the law is often subject to interpretation.

Connectivity

Connecting all of the system components together is an important consideration when planning a camera system. The system designer has numerous options available that can affect the overall operation of the system. Most systems require the cameras to be directly connected to the control equipment with cabling. Some systems or individual cameras can be transmitted back to the control area with wireless transmitters and receivers. Still other cameras may connect to the system via existing cabling connected to a local area network/wide area network (LAN/WAN). Exactly which type of connection is best depends on the system, the application, and the conditions in the facility (see Figure 9.1).

COAXIAL CABLING

Most camera systems are configured using coaxial cable, or coax, to connect all of the components together. Coax is a type of cable that is ideal for carrying an analog video signal over distances that are adequate for most buildings. There are four standard coax sizes that are usually used for camera systems, with the most common being RG/59; the others are RG/175, RG/6, and RG/11 (see Figure 9.2).

Coax cable has one conductor through the center of insulating material called a *dielectric*. This center conductor can be either a solid wire or stranded wire, with a solid conductor being the most common. The center conductor is used to carry the video signal. Wrapped around the dielectric insulation is a wire weaving, which is used as the video signal ground. This weave consists of many small strands of wire interlaced around the entire center insulator and is known as the *shield*. With coax cable for CCTV systems, the center conductor and shield are usually copper or at least mostly copper. This type of cable is also known as 75-ohm cable and is specifically designed to match the 75-ohm impedance-matching requirements of CCTV systems. Other similar cables are available for

Figure 9.1 A rack of video servers is connected to a matrix switcher by coaxial cables and to the network by Cat 5E LAN cables.

other applications and are not interchangeable with CCTV cabling. CATV cabling, for example, is identical except that the center conductor and shield are aluminum. The cables may appear to be identical, but utilizing the aluminum cable in a CCTV system will cause poor picture quality and in some cases no usable picture at all.

RG/175 Coaxial Cable

Also known as mini-coax, RG/175 is the smallest type of coaxial cable used on CCTV systems. This small cable can be beneficial in areas where dozens of cables are run for fairly short distances, because the smaller diameter will help save space (see Figures 9.3 and 9.4).

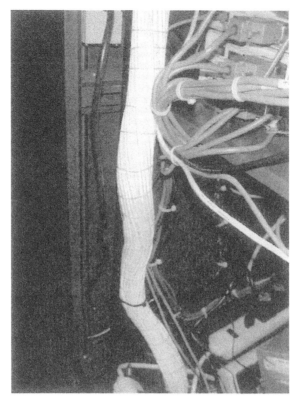

Figure 9.2 Pictured is a bundle of 64 Plenum RG/59 coaxial cables installed in an equipment rack.

RG/175 cable has a center conductor with a wire gauge of 24 AWG. This relatively small conductor means that this cable can only be used for cable runs shorter than 500 feet in length. This type of cable is not a general-purpose cable but is an excellent choice when used properly.

RG/59 Coaxial Cable

Most commercial camera systems connect all of the video equipment together using RG/59 coax cable. For most buildings, RG/59 is capable of carrying the video signal for enough distance for all of the cameras. The actual maximum distance that can be obtained would depend on who is asked. Most cable manufac-

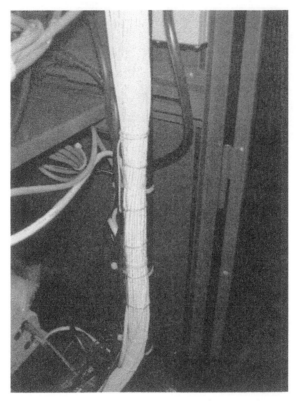

Figure 9.3 Pictured is a bundle of 64 Plenum RG/175 cables installed in an equipment rack. Notice how much smaller and more manageable this bundle is in comparison to the RG/59 bundle shown in Figure 9.2.

turers quote a maximum distance of 750 to 850 feet. Many technicians and product manufacturers claim that 1,000 feet is actually the maximum.

Some devices can extend the maximum distance that can be obtained substantially. Video amplifiers, for example, can be utilized to extend the maximum distance to 1,500 feet or even 3,000 feet. To find such a unit for a camera system, the reader would most likely have to look for a video distribution amplifier, because this is how most manufacturers refer to them. The only real difference is that a true video amplifier only amplifies the signal with a single input and amplified output, whereas the distribution amplifier allows for the video signal to be distributed to multiple locations. Most will have a single input and four amplified outputs. Panasonic manufactures a multiplexer with an optional card,

Figure 9.4 Comparison of RG/59 cable with Bayonet N connectors (BNC) and RG/175 cables with BNC. The top unit is connected with RG/59, whereas the bottom units are connected with RG/175.

which amplifies the video signals and will extend the maximum cable distance with RG/59 to 3,000 feet.

If devices are used to extend the maximum cable distance, the user should know that there are still limits. For example, routing a camera signal through a distribution amplifier to obtain 3,000 feet and then through a second amplifier for an additional 3,000 feet may sound like a great idea, but it will produce poor picture quality in most cases. Because the distribution amplifier amplifies the video signal as well as the noise, a second amplifier will also amplify both. Amplifying the noise twice will be noticeable in the picture quality and may be extremely noticeable when the camera has a low signal-to-noise ratio.

RG/59 cable is available with either a solid-center conductor or a stranded-center conductor. This center conductor has a wire gauge of 20 AWG. Stranded-center conductor coax can only be used effectively with connectors that have a center pin that is crimped or soldered to the center conductor. Twist-on connectors and connectors made with only two pieces will not work properly because the center conductor must be pushed into the center pin of the connector. Solid-center conductor coax is the most common and is the best choice for most applications.

RG/59 is available with PVC casing as well as Plenum, direct burial, and direct sunlight rated. Which type should be used depends on where the cable will be routed and any building code requirements. If the cable will be installed completely in conduit or an area indoors that is not Plenum rated, PVC cabling can be used. If the cable will be run in an area that is Plenum rated, such as a Plenum ceiling, and the cable will not be in conduit, then Plenum cable must be used. To be specific, the cable passing through the Plenum area is the only part that would need to be Plenum rated; the remainder could be PVC. To achieve that, however, would mean that the cable would have to be spliced on both sides of the Plenum area. Some installation companies may do this, because Plenum cable is much more expensive than PVC cable, but it is not recommended.

Regardless of how the splice is done, any splice will reduce the overall maximum distance that can be obtained, and each connection provides an additional point for potential failure. Some companies may ignore the Plenum requirement completely and just run PVC cabling, assuming that no one will ever check. This could be potentially life threatening, however, for the building occupants and should never be considered as acceptable. Plenum cable is required in Plenum areas because of fire regulations. PVC cabling has a lower burn point than Plenum cables and produces toxic fumes. PVC cables installed in a Plenum ceiling means that the toxic fumes would now be directly in the air distribution area of the building. The fumes would spread quickly and perhaps result in loss of life for building occupants. Readers should consult with a building engineer and/or the local fire inspector to determine what requirements exist for the facility.

Riser-rated coax is one other type of cable available in RG/59. Any cable that passes from one floor to the next through a common channel that is known as a *riser* must be at least riser rated. Plenum-rated cable exceeds the requirements of riser ratings, and therefore can be used in place of riser-rated cable. Riser-rated cable cannot, however, be used in place of Plenum-rated cable.

RG/6 Coaxial Cable

Cable runs that are longer than the maximum for RG/59 cable will require a large type of cable to transmit the video effectively. RG/6 is the next size of coaxial cable that is often used with camera systems. It is larger, heavier, and thicker than RG/59 and is capable of carrying the video signal much farther.

RG/6 has a makeup very similar to RG/59 except that it is made of a heavier-gauge wire. The center conductor for RG/6 is 18 AWG instead of 20. This larger-center conductor means that the RG/6 cable can carry the video signal much farther than the RG/59 cable. RG/6 is available in several different options, as is the RG/59, so it is important for the user to know exactly what is required. As with the RG/59, the RG/6 cable is rated as 75-ohm cable. It is also available with a PVC or Plenum-rated outer cover, as well as a riser rated cover. It is available with either a solid- or stranded-center conductor and with a single or double copper shield. RG/6 cable is rated for cable distances of up to 1,500 feet, as opposed to the 750 to 850 feet of the RG/59 cable.

RG/11 Coaxial Cable

When cable runs for the cameras are too long to use RG/6 cable, RG/11 is the next step up if traditional copper coaxial cable is still desired or required. RG/11 is similar to RG/59 and RG/6 but is still larger than the other two. The center conductor for RG/11 cable is a 14 AWG wire, allowing for longer cable runs.

RG/11 is quite large and bulky to use for large camera systems. It can be very cost prohibitive if it is required for use with many cameras. The cable cost alone is much higher than RG/59 and RG/6, and the labor required will also be much higher. Because of the weight of the cabling, shipping charges will be higher, the conduit and/or wire management devices (i.e., cable trays and hangers) will be much more expensive, and the system will be larger and more difficult to install. If the cable runs are long enough to require RG/11 cabling, it may be more cost effective to look for alternative wiring options, such as fiber-optic cable or twisted-pair cabling.

COAXIAL CABLE CONNECTORS

When coaxial cable is used to transmit the video from the cameras to the control equipment, perhaps one of the weakest links is the connector installed at the ends of the cables. Poorly or improperly installed connectors, as well as using the wrong connector type, can cause poor video quality and/or eventual system failure.

Nearly all video equipment used with CCTV systems use a Bayonet-Neil-Concelman connector (BNC) to connect the cable to the equipment. These BNC connectors come in several different styles, each of which connects to the cable in a different manner. The two basic varieties include twist-on connectors and crimp-on or solder-on connectors.

Twist-on Connectors

Twist-on connectors are typically just a single piece that twists onto the coaxial cable. This type of connector may seem like the easiest to install, but it is also the easiest type to install improperly, the easiest type to pull off accidentally, and the most likely to provide poor picture quality.

Twist-on connectors require the center conductor of the cable to be pushed into a tension-type connector in the center of the piece. The coaxial shield is connected to the device by twisting the unit onto the shield and cable sheath. The biggest downfall of using twist-on connectors is the contact between the shield and the connector. If the shield is cut too short, it can make poor or no contact with the BNC connector, resulting in poor picture quality. If the shield is left too long, it either can be torn strand by strand from the twisting action of the installation or can be left exposed from the bottom of the connector. If it is torn, the connection can result in poor video quality; if it is left exposed from the connector, the copper can oxidize over time, which will increase the resistance between the connector and the shield, eventually resulting in poor picture quality or video loss.

Crimp/Solder Connectors

This type of connector is usually available in either two or three pieces. Two-piece connectors require that the center conductor of the coaxial cable be connected to the connector in the same manner as with the twist-on connector. The second piece is then crimped or squeezed onto the shield of the cable. This crimped connection provides a much better electrical contact than the twisting action of the twist-on connector and leaves less room for oxidization.

Three-piece connectors require a small pin to be crimped or soldered onto the center conductor of the cable. When the cable has a solid-center conductor, the connection is usually a crimp connection, whereas with a stranded-center conductor, it is usually soldered.

Crimp/solder connectors provide a much more solid connection to the cable, reducing the chance of a poor connection if installed properly. Many CCTV professionals readily state that twist-on connectors should not be used or should be used only for temporary installations. Still other professionals swear by this type of connector. It is not my intent to determine who is right and who is wrong, but personal experience shows consistent problems when twist-on connectors are used. In nearly every instance where twist-on connectors were installed, the video signal was poor or the system experienced video problems. Also, in nearly every instance, replacing the connectors with crimp connectors resulted in immediate improvement in the video quality.

FIBER-OPTIC CABLING

Fiber-optic cable is an alternative way to send video signals from the cameras to the CCTV control equipment. This method uses lightwaves to transmit the video images and therefore is not limited by voltage loss on a cable. A fiber-optic cable is made of a small glass strand, which is highly polished at both ends. The one-volt video signal is converted to a lightwave signal and sent through the glass fiber to receiving equipment at the other end. There it is converted from a lightwave back to an analog video signal to be connected to the control equipment.

Fiber comes in several configurations, and different types may be required for different camera types. Installing fiber is more difficult than installing traditional coax, and installing the fiber ends requires some level of skill. The ends of the fiber must be highly polished to receive the signal accurately, and if an end is not polished properly, it could mean poor signal quality.

When fiber is used, additional equipment will be required to convert the video signal to light and then back to a video signal. At the camera end of the line, a fiber transmitter or transceiver is required. A coax cable connects from the output of the camera to the input of the transmitter. The fiber then connects to the output of the transmitter for transmission to the control equipment. At the control equipment end of the fiber, a receiver or transceiver is required for the conversion back to a video signal. The fiber connects to the input of the receiver, and a coaxial cable connects from the output of the receiver to the input of the control equipment, such as a multiplexer or matrix switcher (see Figure 9.5).

Drawbacks to using fiber include the cost of the fiber and additional equipment, as well as the additional labor cost. Labor cost will be increased because of the time required to install high-quality fiber ends. Much more care is also required when installing fiber-optic cables to avoid breaking the glass fiber. Although the pull strength of fiber-optic cable is not necessarily lower than that of coaxial cable, it is much easier to break a thin strand of glass than it is to break

Figure 9.5 The Kalatel Digiplex system pictured here includes a fiber-optic module for easy connection to the control portion of the system. In the foreground are a few fiber-optic cables with typical connectors.

a thicker core of copper. Repairing a broken fiber-optic cable is extremely difficult if not impossible.

One alternative to installing traditional fiber is to install modular fiber that is preconfigured for the correct length and with the proper connectors. Although the initial cost may be more than that of traditional fiber, the overall cost may actually be lower. Modular fiber has the fiber ends installed and tested at the factory to ensure that it is properly installed and polished. Usually the ends are part of a separate piece that plugs into a special adapter installed on the fiber. This way the ends are not pulled when the fiber is installed. With traditional fiber there is a maximum bend radius of 6 inches or more. This means that the fiber cannot be bent at a sharper radius than that or it will break. Some modular fiber, on the other hand, has a bend radius as small as 1 inch, making it much less prone to damage. When comparing the cost of traditional fiber with the cost of modular fiber, the cost of all equipment, fiber ends, fiber, and labor should be compared to see which will be more cost effective for a particular application.

Fiber-Optic Trunks

When traditional coaxial cable is used, each cable is usually run continuously from the camera to the control equipment. When fiber is used, there is an alternative that can be beneficial. If the cameras are run to a secondary location that is closer to the group of cameras, they can be connected to what is known as a *fiber trunk*. This fiber trunk would then run to the control equipment for connection to the fiber receivers.

Fiber-optic cable is available with different quantities of fiber strands within an individual cable. Standard configurations include single strand, 2 strand, 6 strand, and 12 strand. If 12 cameras are located in one general area that is a substantial distance from the control equipment, each camera could be routed to a central location with single- or two-strand fiber. These fibers could then connect to a larger fiber, such as a 12-strand, so that only a single cable would need to be run back to the control equipment, instead of 12 individual cables. This would present cost savings in the amount of fiber required as well as the labor required for installation.

This is particularly true when multiple cameras are located in one building and must be transmitted to another building. Running a single trunk cable would be much more efficient than attempting to route all of the cameras back to the control equipment individually. It is usually a good idea when using a trunk return to make sure that plenty of spare fiber strands are included. In many installations the designer will call for double the necessary fiber strands to allow room for future expansion and in case any fiber strands become damaged.

If an installation is highly critical or very high security, it may also be beneficial to run two separate trunks along two different paths in case one trunk cable is ever damaged or compromised. In this way, the entire system could be restored to operational by switching from the damaged trunk to the backup trunk rather easily.

Transmitters and Receivers

Fiber-optic transmitters and receivers come in several different varieties for different functions. If a camera connected to fiber has pan/tilt and zoom (PTZ) capabilities, the transmitter and receiver must be capable of converting the control commands as well as the video signal. Usually these units are larger in size than transmitters and receivers for standard fixed cameras, as well as more expensive.

If fiber transmitters are required, it is important to select units with a power-input requirement that is compatible with the camera locations. If the camera uses 12 volts DC and the fiber transmitter is 24 volts AC, a second power supply would be required so that both voltages are available at the camera location. The system designer should make sure this requirement is considered when selecting equipment. If the transmitter will be powered from the same voltage source as the camera, it is also important to ensure that the total current draw is calculated for all of the powered devices.

Controlling PTZ Cameras

When controlling PTZ cameras that are using fiber-optic cable, it is important to choose the proper transmitter and receiver. To know which one is correct, the designer must know what type of signal is being used to control the camera. Whether the camera is using a coaxial signal transmission or additional cabling through a 232 or 485 port, fiber transmitters and receivers can be purchased that will meet these needs without running copper cabling to the camera location. If the camera control uses a separate control cable, it may be necessary to use two fiber strands for each camera. One strand would be used for the video signal, while the second strand would be data transmission for control of the PTZ functions.

Just as PTZ cameras are more expensive than standard cameras, fiber transmitters and receivers for PTZ cameras are more expensive than basic transmitters and receivers. They are also larger than units for fixed cameras, so it is important to allow for enough space during the setup and design.

TWISTED-PAIR CABLING

Another alternative to coaxial cable for video transmission is the use of twisted-pair cables. The principle behind this type of cabling is similar to that of fiber-optic cable. Transmitters and receivers are used to convert the signal for easier transmission through a twisted pair of copper wires. Unlike fiber, however, no special connectors are required to connect the equipment together. This means that a tremendous amount of time will be saved during the installation, along with the lower cost of the cabling. With twisted-pair technology, practically any pair of unshielded twisted-pair wires will work effectively to send the signal from one end to the other. Wire size is of little concern, and cable distances are high, depending on the types of transmitters and receivers chosen.

Passive Transmitters and Receivers

Passive transmitters and receivers are small units that require no external power source. The video output of the camera is connected to the input of a transmitter through a BNC connector. Two screw-down connectors are then used from the video output to connect the transmitter to the twisted-pair cable. At the other end, a similar setup is used to connect the twisted-pair cable to the receiver and the receiver to the control equipment.

Passive transmitters are usually used for camera locations that are 1,000 feet or less from the camera control equipment. Any distance greater than 1,000 feet will have a signal loss that could result in either poor picture quality or no picture at all. If the camera will be located at a farther distance, active transmitters and receivers would be required.

Active Transmitters and Receivers

Active transmitters and receivers for twisted-pair cabling are used when the distance from the camera to the control equipment is too great for passive units. Active units are more expensive than passive units, and they usually require an external power source.

Active units are available in a few different styles that vary based on the distance required. Basic active units usually have a maximum cable distance of 1,500 to 3,000 feet. They will go up in price as well as the maximum allowable distance from there. Some units are made that can be effectively used for cameras located at distances of up to 3 miles from the control equipment location, and some much farther than that.

One primary reason to use twisted-pair cabling instead of coax or fiber is the cost factor. Twisted-pair cable, even in Plenum, is substantially cheaper than any coax or fiber-optic cable. Typically, for new installations, designers will specify that either Cat 3 or Cat 5 network cabling be utilized. Both of these cable types contain four twisted pairs of wire. Because only one twisted pair is required for the video signal, the other three pairs can easily be used to provide power to the camera. If this technique is used, the designer should ensure that the voltage drop will not be too great to prevent the camera from working properly.

Another benefit to twisted-pair cabling is that existing cabling in a building can often be used. Network cabling can easily be used for video signal transmission, as can standard telephone cabling. When the telephone is installed, there are often dozens of spare pairs run throughout a facility. For example, a 50-pair trunk may run from one telephone closet to another, yet the phone system may only utilize 15 pairs. For existing buildings in which it may be difficult to run new cables between floors, this can mean a tremendous amount of time saved as well as reduced equipment costs.

Alternate Configuration Possibilities

As with fiber-optic cable, twisted-pair systems can use trunk cabling for routing multiple camera signals from a remote location. This gives the designer the option of running a multiple-pair cable from one area to the control equipment area instead of routing every camera all the way back. A single 25-pair cable could be used, for example, instead of 25 individual pairs, making cable management much easier. If this technique is chosen, it may also be more feasible to locate the camera power supplies closer to the camera locations. If the power supplies are located at the camera end of the trunk, then the power can be distributed from there, meaning less of a voltage drop for each camera. Rather than splicing each twisted-pair cable to a pair from the trunk, a patch panel or even telephone punch-down blocks, such as a 66-block, can be used. An added benefit of this system is that it provides the technician with an additional point for troubleshooting.

NETWORK VIDEO CABLING

In the modern digital age, installing cameras in a building has become even easier. Many cameras are now available that connect directly to the LAN/WAN network of any office or facility. These cameras do not require dedicated cabling from the camera to a monitor. All that is required is a connection to any network port within the building (see Figure 9.6).

These cameras each have their own unique network address known as a Transmission Control Protocol/Internet Protocol (TCP/IP) address. On the back of the camera is a connector that looks identical to the type found on office walls in which computers are usually connected. A cable is connected from this camera connector to any open network connector in the room, and the video is accessed by simply entering the TCP/IP address from any Web browser.

If the camera is installed in an area that has no empty network ports, a hub can be installed near one of the devices that is using a port, such as a workstation. The workstation and the camera can then both be connected to the hub to use the same network line.

Figure 9.6 Each of these camera servers is connected to the computer network with Cat 5E LAN cables. Each server has four camera inputs and allows for each camera to be accessed via the network.

There are, however, several disadvantages to using network cameras with a traditional camera system. Most important, if the camera output is to be recorded or viewed through a traditional multiplexer, the video signal must be converted back to an analog signal first. Most multiplexers and recorders used for security camera systems do not have a network connection for camera inputs and only allow for a standard analog video signal input.

Another disadvantage is that the video signal will use bandwidth from the network. Although this may not be significantly noticeable with a single camera on many systems, in some cases it can be a strain on the network. If multiple cameras are used and the network is relatively small or limited in capacity, the cameras can slow down the overall communications for the entire network.

If network cameras are used, it is important to make sure that the network people, such as the information technology (IT) department, are involved throughout every step. They should be involved in the purchasing decision as well as the implementation to ensure that the cameras are compatible with the network. Failure to involve the IT staff from the beginning could cause major conflicts with the network, the camera system, and the general work environment that could easily be avoided.

WIRELESS VIDEO TRANSMISSION

So far this chapter has looked at the different types of cabling options for connecting cameras to the camera system. One option that has not been addressed yet is using no cabling at all. Many cameras are now manufactured with wireless video transmitters built into them. This means that the video signal is sent by a predetermined frequency to a receiver, which then converts the signal back to an analog video signal.

There are several advantages and disadvantages to wireless video transmission when used with a traditional camera system. Each may or may not be a factor on a system application and should be understood before deciding whether to use a wireless camera.

One distinct advantage of a wireless camera is the ease of installation. A wireless camera can easily be installed in a location that would normally be difficult to cover because of cabling difficulties. In areas that are either decorative or would be difficult to run a video cable to, the wireless camera could be installed. This is not always as much of an advantage as it would seem to be, however. Even though no video cable is needed for the wireless camera, it still requires power, which is another cable. It may be possible to get power for the camera locally, but, if not, a power cable is required anyway. If a power cable can be installed, then a video cable can probably be installed as well.

Wireless transmitters are available for a wide range of video transmission distances, which is another advantage. With the proper wireless transmitter and receiver, it is possible to receive the video from a camera at a location as far as 15 miles away. For long distances like that, traditional cabling options are not practical or feasible.

With wireless video signals, it is possible to receive the video at multiple locations, with multiple receivers on the same frequency. A single wireless camera could easily be viewed in several locations by simply adding receivers at the viewing locations. This ability to view the video with additional receivers can also be a disadvantage, however. Because the video signal is no longer part of a closed circuit, anyone with the proper receiver can tune in and view the video signal. Because most off-the-shelf wireless transmitters and receivers are available in only two different frequency ranges, it would not exactly be difficult to figure out what frequency was required if someone wanted to tap into the video signal. There would be no indication at the authorized receiving location that the video was being picked up somewhere else.

A primary disadvantage of wireless cameras is the limited number of frequencies available for off-the-shelf units. Most commercially available wireless cameras or transmitter/receiver combinations are available in either the 900MHz or 2.4GHz frequency ranges. Within those ranges, most manufacturers are limited to four frequency channels. This means that the camera system is limited to a maximum of four wireless cameras per system. More could be achieved by using a combination of both frequency ranges and by meeting licensing requirements for amateur radio operators.

With long-range transmitters or receivers, the limited number of frequencies becomes more evident, particularly with units having a range of several miles or more. With these longer-range units, any transmitters from other organizations could potentially interfere with the desired signal, meaning that the user does not get a good, high-quality video signal. It could also mean that the receiver picks up the wrong video signal or the receiver of the other system owner picks up the signal. Most of these long-range units are directional and must be aligned for the best signal, which does help limit the possibility of interference, but the possibility is still there.

900MHz Systems

Commercially available video transmitters and receivers are most commonly available in two frequency ranges. The first frequency range is the 900MHz band. This is the same frequency range used by many cordless telephone systems and is reserved primarily for low-power output devices.

Video transmitters utilizing 900MHz typically have a low power output and a low reception range, meaning that the video signal cannot be picked up miles away. With so many 900MHz items available and in use, if the transmitters did have a long range, they would probably have an excessive amount of interference anyway. Most 900MHz receivers have the ability to receive four different video signals or channels. Each is selected one at a time from a pushbutton. To view all four images simultaneously would require four separate receivers.

An advantage to the 900MHz transmitters and receivers is that they are readily available. This is also a disadvantage, however, because interference and duplication of frequency are possible, as is signal interception. Another advantage is that equipment cost is relatively low, and no FCC operator's license is required for low-power units.

2.4GHz Systems

Transmitters and receivers in the 2.4GHz range are relatively new when compared with the 900MHz units. The 2.4GHz units operate at this higher frequency and are more efficient than the lower-frequency units. At the higher frequency, the video signal is less prone to interference and easier to receive at longer distances. Some 2.4GHz units are capable of being received up to 15 miles away.

Advantages and disadvantages of the 2.4GHz transmitters and receivers are similar to the 900MHz units, except for the reception range. They are both readily available, with the capability of receiving four video signals, and have a reasonable cost. Most 2.4GHz units do not require an operator's license and do not have any special restrictions.

MICROWAVE LINKS

Another alternative for longer-range video transmission is *microwave linking*. This type of system uses a microwave transmitter and receiver dish to transfer the video signal from one location to another. The video transmitters and receivers are usually mounted up high, such as on the roof of a building, and must be aligned to work properly. Microwave transmission only works by line of sight, which means that the transmitter and receiver must be aimed directly at each other. This requires some coordination at both the transmitting and receiving site to ensure that the signal is being received properly.

Microwave systems are most prone to problems when new construction goes up between the transmitter and receiver, and occasionally during heavy storms. The signal will also be lost if either the transmitter or receiver is bumped or moved and must then be realigned. In addition, in some cases microwave systems may require FCC licensing before purchase or installation.

CONCLUSION

Getting the video signal from the camera to the other equipment is often taken for granted. Most often it is assumed that coaxial cable such as RG/59 will be used, and alternatives are not considered. Just as with choosing any other component of the camera system, the video transmission medium should be carefully considered.

Although coaxial cable is always an excellent means of getting the video signal from one point to another, it may not always be the most efficient or cost-effective way to wire the system. For longer distances within a building, it may be more cost effective in some instances to use fiber-optic cable or twisted-pair cables.

When deciding on the type of transmission media to be used in a system, it is important to look at all aspects and associated costs of using that particular medium. This would include the cost of the cabling; the amount and cost of the labor to put it in; the cost and labor required for the connectors; and the cost and labor for any transmitters, receivers, or converters required. When all of these factors are considered, as well as the video quality from the selected medium, the designer is much more likely to select the best transmission medium for the job.

10

Outdoor Considerations

So far this book has mainly looked at elements and components of a camera system that are used within a facility. When cameras are required on the exterior of the facility, many additional considerations must be taken into account to properly design and set up the system.

Many of these are the same considerations for choosing indoor equipment, but the influences are different. If the outdoor considerations are not considered separately, it could result in a negative impact on the entire camera system, from ease of use to the amount of maintenance and repair required.

Climate or environmental conditions, lighting variations, and equipment accessibility are the three major areas that must be considered in detail when using cameras outside. This chapter looks more closely at all three of those factors to help you understand the best way to select outdoor equipment and locations.

CLIMATE

Climate throughout the world can have a dramatic effect on what equipment will work properly and what will cause a system failure. Just as with winter coats, camera enclosures with a heating element are not a big requirement in Honolulu. Every area of the world has its own unique weather characteristics that affect equipment choices. Temperature, humidity, rainfall, likelihood of electrical storms, and even air quality can influence the choice of outdoor camera equipment (see Figure 10.1).

Harsh Environments

Typical environmental conditions for much of the world can be marked by four seasons, with standard variations in temperature and humidity. Temperatures

Figure 10.1   Installing cameras outdoors adds more variables to consider when selecting CCTV equipment.

never really get excessively high or low, and most camera equipment can be adapted to the local conditions. Some areas of the world do not fall into this category, however, and improper equipment choice can result in rapid equipment failure. Arctic regions, for example, typically have very low temperatures for much of the year. If the camera equipment is not configured to operate in an extremely cold environment, it will not last long. Desert regions, on the other hand, rarely if ever get below freezing in most of the world, so the same setup would not work.

Although this observation may seem like common sense, it is often overlooked when designing a camera system. When the climate is considered, the different ways that the camera equipment can be affected are often not fully assessed.

Cold Environments

When outdoor cameras are going to be installed in areas that may experience below-freezing temperatures, it is important to ensure that the camera will always stay within the manufacturer's operating temperature range. This goal is usually accomplished by installing the camera in an environmental housing. Heating elements would then be used within the housing to keep the internal temperature above the camera's minimum temperature.

A sensor within the housing will detect when the temperature falls to a certain threshold. This will cause the heating elements to turn on and heat up the area directly surrounding the camera and lens. Once the temperature within the housing increases to the high-point threshold, the heating elements will turn off. The high-point threshold is usually several degrees above the temperature that turns the heating elements on. In this way, the temperature variation is never very large, and the heating elements can shut off when they are no longer needed.

Some climates are cold enough that the heating elements will remain on for long periods to try to keep the camera housing warm enough for proper operation. For extremely cold environments, it might be necessary to have a larger quantity of heating elements to keep the system working properly.

When an outdoor camera requires heating elements, it is important to remember that during these cold seasons the power requirements for each camera location will be much greater. Every time the heating elements turn on, this will cause a substantial increase in the current draw on the power cables. The cabling and power source must both be sized accordingly to allow for this current increase requirement.

An indoor pan/tilt and zoom (PTZ) camera, for example, could typically require 30 volt-amps for proper operation, whereas an outdoor PTZ with heaters could require 90 volt-amps. This means larger cables to carry the power from the supply to the camera, and a larger supply capable of delivering the additional current would be needed.

Hot Environments

Hot climates can be difficult for designing camera systems as well. If cameras are to be used outside in environments that can have very high temperatures, it will be important to make sure that the camera equipment stays cool enough to operate properly.

When an outdoor camera is to be used in a hot climate, it is usually equipped with cooling fans that work in a manner similar to heating elements. A temperature sensor triggers the fans to turn on when a certain temperature inside of the camera enclosure is reached. These fans then blow air across the camera and lens to help cool them down. Many manufacturers and technicians refer to these fans as blowers. When a technician talks about a camera housing with heaters and blowers, that means that the housing is equipped to keep the camera warm in the cold months and cool in the hot months.

Just adding cooling fans and a temperature sensor to a camera enclosure is not always sufficient to make sure that the housing temperature is low enough for proper camera operation. If the camera housing is in direct sunlight in a very hot environment, the cooling fans often cannot cool the housing down enough. In this type of environment, it may also be important to shield the camera housing from direct sunlight to help keep the internal temperature down. Therefore, many manufacturers have an optional sunshield for camera housings. The shield usually covers the top of the housing and has a gap between itself and the housing for airflow.

If the climate consistently has high temperatures and direct sunlight, it would be wise to protect the camera housing as much as possible from prolonged direct sunlight. This would be in desert and tropical environments with consistent temperatures over 120°F. For these environments, placing the camera housing under a protected eave or in a location that is shielded at least partially from direct exposure will help keep the temperature within the camera housing down and allow the cooling fans to work more effectively.

Saltwater Environments

Perhaps one of the harshest environments for equipment is a saltwater environment. In areas close to large bodies of saltwater, such as oceans and seas, the salt content of the air can be very high. This salty air can be corrosive and can have long-term negative effects on the camera equipment (see Figure 10.2).

If the camera housings, cables, and other exposed components are not protected or designed to work in a saltwater environment, the components can actually be eaten through by the corrosive exposure. Pitting and rusting of metal pieces and buildup of salt residue can destroy the housings, cables, camera mounts, and pan/tilt units.

Many manufacturers make equipment with a type of paint or protective layer to help combat these negative effects, but the equipment will still require more maintenance than in an environment without saltwater. Housings and equipment should be visually inspected periodically for signs of corrosion and wear. If any blistering is apparent on the paint or if any corrosion has begun, it should be cleaned and covered as quickly as possible to prolong the life expectancy of the equipment.

Cameras installed in areas around pools are often subject to a similar corrosive environment. Indoor pools, for example, often have a fairly high content of chlorine in the air, which is also corrosive. Long-term exposure can shorten the life span of the equipment if it is not inspected and protected frequently.

Seasonal Considerations

Extreme climates are not the only cases in which environmental conditions should be considered when designing the camera system. Much of the world has

Figure 10.2 Saltwater environments can shorten the life span of outdoor cameras. Pictured here is an outdoor dome camera installed on a light pole in Virginia Beach.

climates with varying conditions throughout the year. Many areas experience temperatures well below freezing in the winter and temperatures over 100°F in the summer. In these environments, outdoor cameras would need to be equipped with heating elements and cooling fans for year-round protection.

Temperature is not the only concern when looking at climate and weather conditions. Humidity extremes can also cause camera equipment to be damaged. Particularly in the winter months, humidity can be very low. This low humidity can leave electronic equipment prone to damage from electrostatic shock or static electricity. Good system grounding is essential to help protect equipment from electrostatic shock. Technicians and anyone who may come into contact with the equipment must also take great care to ensure that they do not transfer a static charge to the camera equipment when they go to work on it. Properly grounding themselves and the equipment only takes a second and can help ensure that the equipment is not damaged.

High humidity could also cause a water vapor buildup inside of the camera housing. The only real way to avoid this problem is to make sure the housing is clean and dry at all times. This buildup can fog the glass or the lens, which is usually an indicator that someone should physically check out the camera and housing.

One way to avoid potential problems from high or low humidity is to use an environmentally sealed and pressurized camera assembly. This type of assembly usually consists of a camera and lens inside of a special housing equipped with heaters and cooling fans. The housing is usually completely sealed and filled with nitrogen gas at a pressure of about 15 pounds per square inch. If a pressurized housing is used, it is important to remember that if any work is required inside the housing, it must be resealed and repressurized or it will be ineffective for long-term protection.

LIGHTING

Lighting is one of the most talked about issues when discussing outdoor cameras. There are many different types of lighting, all of which can have different effects on picture quality or whether a picture can be seen at all. Many books have discussed lighting with camera systems and which types of lighting are best, but the system user most often has no choice when it comes to types of lighting. Even in new buildings, lighting is often not thought out for the effect it has on camera pictures; rather, it is usually chosen based on the cost effectiveness of the lighting type and the esthetic aspects it will have on the building and surrounding areas.

With indoor cameras, the light levels usually depend on light fixtures even in the middle of the day, unless the sunlight through windows affects the levels. Many interior cameras, such as in hallways and basements, never see anything except artificial lighting, so the light level is usually constant.

With outdoor cameras, the light levels are much broader than they usually are with indoor cameras; the variation can be from a bright sunny day to no light at all. For this situation the camera choice can be much more relevant than with indoor cameras when it comes to lighting. The difference between black-and-white cameras and color cameras also becomes more evident. As mentioned earlier in this book, color cameras do not operate as well under low light conditions as do black-and-white cameras.

During the day the light level from the sun can be as high as 100,000 Lux during the sun's highest point. At night, however, the natural light is low enough so that artificial lighting is required for most cameras. Nighttime is not the only time that light levels are a concern, though. Even during the day when there is a sufficient amount of light for a good camera picture, the nature of the light can cause camera problems.

Low Light Levels

Probably the most obvious potential problem with lighting is the lack of sufficient natural light at night. In the evening, artificial light is required to ensure that enough light is available for an adequate camera picture. If the cameras are color, the amount of light required will be noticeably greater than that needed for black-and-white cameras. The type of artificial light that is available can have a tremendous effect on the quality of the picture.

In many cases, the amount of light required for good picture quality may not be feasible for use of a color camera. Some localities may even have restrictions on the amount of light that can be used and the types of lights that can be used. Businesses on the approach paths to airports, for example, may be required to keep the light level below a certain level and to use only lighting that is obviously different from that used by the airport.

If the amount of artificial light available is not adequate for a color camera, it may be adequate for a decent-quality black-and-white camera. If the camera is going to be installed at an existing facility, the light levels can be easily measured with a light meter. Light meters are readily available at any good photography store and are relatively inexpensive. A good-quality light meter can be pur-

Figure 10.3 Multiple light sources in an otherwise dark area can cause real problems when setting up outdoor cameras. This picture of the Luxor in Las Vegas shows an extreme example of how lights can affect the picture.

chased for less than $200 and will display both foot-candles and Lux levels. If the lighting available is provided from sodium vapor lights, make sure that the light meter can detect this type of light. Many lower-cost models cannot detect light from a sodium vapor light source accurately.

To accurately measure the light levels with a light meter, the measurements should be taken in all areas that may be part of the viewing subject, not just at the location of the camera. Light levels will vary from one part of the viewing area to another, so several measurements should be taken. Lighting that puts out decent light levels but in a small area can actually cause more problems than lights with lower levels spread over a wider area (see Figure 10.3). A camera that is focused on the area with a higher level will adjust for that level, and the areas with less light will appear quite dark. If the light levels are lower but fairly even, the camera will adjust to the level in the main focal area but will still be somewhat consistent within the entire viewing area.

Light levels should be taken before a camera choice is made. Once all of the measurements are done for each camera location and viewing subject, compare the numbers to the minimum illumination requirements in the camera specifications. This should help narrow the number of camera choices and considerably improve the anticipated picture quality.

If the light levels available are not sufficient to provide a good picture quality, it may be necessary to increase the light levels. If it is not possible to add more lights to the viewing area, infrared illuminators can be used with black-and-white cameras to help with the picture quality. Black-and-white cameras have good sensitivity to light in the infrared range. The type of infrared illuminator to use will depend on the area to be covered and the camera application.

There are two primary types of halogen infrared illuminators currently used with camera systems, both described by the wavelength of the infrared light that they transmit. The first type is a 715-nml illuminator. This type transmits infrared light from 715 nm and higher. The human eye can see light in the infrared spectrum only up to about 780 nm, with sensitivity to light higher than 700 nm being weak. This means that the human eye will detect the light coming out of this type of illuminator, but it will appear weak and red.

The second type of infrared illuminator transmits energy at 830 nm and higher. All of this infrared light is outside the capabilities of the human eye, so the light being transmitted will be undetectable to people. This type of illuminator is more ideal for applications that require completely hidden surveillance at night, because the 715 nm unit would show a red glow.

One primary shortcoming of halogen infrared illuminators is the short life span they have. These units produce a large amount of heat, much of which will remain trapped inside of the light. This heat causes the light to have a short life cycle of only 1,000 to 2,000 hours of operation. Because of this short life span, new types of illuminators were developed that have become much more common with camera systems.

Many new infrared illuminators use light-emitting diodes (LEDs) to transmit infrared light. The amount of light produced from a single LED is small and would not provide much help for the camera. To compensate for this LED, illuminators use a large group of LEDs usually arranged in a grid or matrix pattern. The combined infrared energy from the group of LEDs will produce enough infrared light to enhance the view of the camera.

LED illuminators are usually described by their wavelength, their power output (usually 7, 15, or 50 watts), and their radiation angle, usually between 30 and 40 degrees. This viewing angle describes the width of projection that the infrared energy will be transmitted from the LED, so a 30-degree angle will produce a much more narrow coverage pattern than a 40-degree angle.

The primary advantage of using an LED illuminator is the long life expectancy. Instead of lasting only 1,000 to 2,000 hours, LEDs can last up to 100,000 hours. That means that a single LED illuminator could last as long as 100 consecutive halogen illuminators. This is much more cost effective, not only from an equipment standpoint but also from a maintenance and labor standpoint.

Artificial Light Sources

For camera coverage at night, a wide variety of artificial light sources can be used. Which type is available and which type is best for a particular application will have a great impact on the quality of the images produced at night. Artificial light sources fall into three main groups, depending on the wavelengths they produce.

The first main group is those lights that produce incandescent light. Candles, halogen lamps, tungsten lamps, and standard lightbulbs all fall into this category. These lights all provide a smooth and continuous light spectrum, making them work well for color and black-and-white cameras. This is to say that the light they produce covers a broad range of wavelengths, both visible and invisible to the human eye. A large portion of the lightwaves they produce fall well into the range that both color and black-and-white cameras can detect. It is not always feasible, however, to have enough of these light sources outside to produce the amount of light required by the cameras.

The second main group is those lights that produce light created by an electrical discharge through a gas or a vapor. Neon lights fall into this category, as do sodium vapor and mercury vapor lamps. The wavelength produced by this group of lights depends on the type of gas or vapor that the electrical charge is dissipated through. This can be distinguished with the human eye and is much more apparent through a camera. A mercury vapor light, for example, will have almost a bluish-white appearance, whereas a sodium vapor light has more of a yellowish tint.

Mercury vapor lights cover the violet to greenish-yellow spectrum of visible light, centering mainly around blue. Areas that are lit with mercury vapor lights will appear to be a bluish-white hue. Black-and-white cameras will usually have a decent image displayed. Color cameras will display a unique display with greens, yellows, and blues appearing to have a blue tint and reds having a sharp and bright contrast. The display for color cameras will appear to be substantially better than with sodium vapor lights, but it still will not look like the traditional color image. Sodium vapor lights produce a light that covers what the human eye sees as green, yellow, orange, and red, with yellow being in the center range of the spectrum. Sodium vapor lights used with black-and-white cameras show a decent-quality black-and-white image. When used with color cameras, the image will appear to be flat and monotone. Colors will not be sharp and will appear almost as various shades of brown and yellow. This is still better than no picture at all, however, and should not necessarily be considered a bad thing, just not the ideal color picture.

The third main group of artificial light types is fluorescent lights. With fluorescent lights, a gas discharge in a tube emits a visible ultraviolet light, which causes a phosphor coating on the inside of the tube to glow. Fluorescent lights cover the full visible light range of the mercury vapor, and the sodium vapor lights combine plus a little more. It has a more continuous and even spectrum coverage than the sodium and mercury vapor lights but still has some higher outputs of particular wavelengths. Although the visible light coverage is higher than the sodium and mercury vapor, it is still not near that of the tungsten lights.

Color images viewed with fluorescent lights will have fairly sharp color contrast, as can be seen with most indoor cameras, but will appear slightly different from the same scene viewed with natural light. Color temperatures and the white balance of the cameras should be observed, if possible, if fluorescent lighting will be used.

Day/Night Cameras

Because of the available lighting, it is sometimes not feasible to have color cameras at night. This means that black-and-white cameras must be used for the desired night coverage. During the day, however, it may be important or desirable to have color images of the same scene. If this is the case, then a day/night camera should be considered. A day/night camera displays color images during the normal daylight hours and automatically switches to a black-and-white image for coverage when the lighting is insufficient for color. This type of camera can provide the most desirable image as often as possible, making sure that the coverage area is always adequately visible on the camera system.

Day/night cameras seemed to come about because of the nature of traditional color cameras. Color cameras have what is known as an *infrared cut filter,* which prevents the infrared light energy from affecting the picture. That is the primary reason that a black-and-white camera will have a better image at night, because the color camera by design has less light to work with. This is not the exact scientific reason for the differences between color and black-and-white cameras, but it is the best way I know of to put it into basic terms.

The ability of a camera to automatically change from color imaging to black-and-white imaging and back is done primarily in two different ways. The first way used by many manufacturers is to remove the infrared cut filter when the light levels get to a certain level. This allows the sensing chip inside the camera to utilize the available infrared light, producing a black-and-white image. Another method that some manufacturers use is to build a camera that actually contains two separate sets of imaging circuits, or two chips. During the day, the color chip is used for the picture, and at night the black and white chip is used. Both methods are effective and are much more cost effective than installing dual cameras for the same coverage area.

To prevent a day/night camera from toggling back and forth from black-and-white to color, a built-in light sensor detects when the available light falls below a certain threshold. Once the light available falls below that threshold and stays below it for a predetermined period, the camera will switch. This time delay ensures that the camera does not toggle back and forth because of a passing cloud, overcast conditions, or a bright direct light, such as from a car's headlights.

Day/night cameras are slightly more expensive than a standard color camera but not by much. Early day/night cameras could cost thousands of dollars, but as the demand grew and production volumes increased, the prices have dropped to affordable levels. In my opinion, the value of this type of camera far outweighs the cost.

Direct Sunlight

So far this chapter has focused on the effects of insufficient and artificial light on video quality. The bright light in the middle of the day, however, can also have a tremendous impact on camera choice and picture quality.

As the demand for smaller and smaller cameras grows, the size of the image-sensing chip has shrunk. One side effect from the decreasing chip set is the effect that light has on the image, particularly with less expensive cameras. With 1/3-inch and 1/4-inch chips, for example, a common image phenomenon is the starburst effect or the appearance of bright white lines vertically on the screen. A bright spot within the image usually causes this when most of the rest of the image is fairly dark. It can also be seen, however, when an object such as a person comes between a bright source of light and the camera, known as *backlighting*.

With many of the less expensive cameras, there is no way to avoid this image flaw. Proper camera choice initially, however, can eliminate it in most situations. For most outdoor camera applications, it is likely that there will be periods of bright light, backlight problems, and bright reflections that will affect the image quality. If that is the case, it is important to look at the various options available on the higher-end cameras. Digital signal processing (DSP) is one such feature that can help ensure a decent picture quality even in the most adverse lighting conditions. If frequent backlighting is the only problem, then a camera with backlight compensation may be sufficient to provide a decent picture quality.

Camera location can also help or hurt the picture quality. If the camera must look almost directly at the sun during sunrise or sunset, not only will picture quality be adversely affected, but the camera may also not last nearly as long as it could have. If it is possible to have the camera in a higher location looking down at the scene below, it may be less likely that the camera will at any time be aimed directly at the sun. The scene to be viewed, conditions throughout the day, and seasonal conditions should all be considered when deciding on the permanent location of any outdoor camera.

ACCESSIBILITY

Camera location brings up the final important factor for this chapter when considering outdoor cameras. Accessibility of the camera equipment can be important and is not often considered until it is too late. A camera is typically installed somewhere that will give the best view of the desired scene and where it is least likely to be vandalized or tampered with. Although these factors are important to consider, this approach often leads to cameras being placed in locations where they are extremely difficult to install and even more difficult to service.

When planning for outdoor cameras, it is important for the designer and technician alike to consider how the camera location will affect the installation labor and long-term serviceability of that camera. If installing the camera initially will require scaffolding or a bucket lift, it will require the same thing any time that camera needs to be cleaned or repaired in the future. Before deciding on that location as the final choice, the designer and technicians should look at any viable alternatives that still provide the desired coverage, are still protected from tampering, and provide for better accessibility in the future. The phrase "what goes up must come down" also applies to camera installations. At some point in the future, something at that camera location will need to be cleaned, repaired, or upgraded.

Installation Access

If the technicians are saying to themselves or aloud, "I'm glad I'll never have to get up here again," chances are the location is a poor choice for a camera and alternatives should be considered. Outdoor cameras are typically installed in a few standard locations. Wall-mounts are common and can be appropriate for cameras that need to be placed at fairly low heights on a building. The advantages to a wall-mount camera are the relative ease of installation and cabling and, if installed properly, the ease of future service. These points only hold true if the cameras are installed at a fairly reasonable height, however, such as at a point accessible from a standard installation ladder. Wall-mount cameras can be impractical when the installation height becomes excessive.

Installing a wall-mount camera 40 feet up on the side of a building may cause installation and service problems, for example. This is too high for most ladder installations and too high for most small scissors lifts. That means that the installation and future service technicians would need to use either a larger lift or scaffolding, or put themselves at risk by using excessively long extension ladders. All of these choices will increase the cost of installation and service as well as put the technicians at more risk unnecessarily. Unless there is a specific reason that the camera must be in that particular location at that height, safer and more convenient alternatives should be considered.

Pole-Mount Cameras

Another typical mounting method is to pole mount the camera. There are several ways in which a camera assembly can be done with pole mounting, from independent camera poles to placement on existing poles. Each has benefits and drawbacks that must be considered.

Mounting cameras on existing poles means that new poles will not have to be installed for the cameras. The existing poles are usually a light pole, which means that the camera will have some light to work with at night. If the lights are the same voltage as the camera requirements, such as 110/120 volts, it may be more practical for an electrician to help access power directly at the pole.

A drawback to using existing light poles is the power, however. The power for the outdoor lighting is usually not the same voltage as the regular building circuits. Lighting power could be 277 volts, for example, which is not practical for use by the camera system. This also brings up another situation, where the technician might be placed in harm's way unnecessarily. To install a camera on the existing pole means that video and power wires must be routed to the camera. If the cabling is to be routed inside the pole, it will be in proximity to the power cables for the lights, placing the technician dangerously close to high voltage. This is a risk that the technician must often assume as an everyday part of the job, but if this risk can be avoided, it should. Running the video cables too closely to the high-voltage cables can also induce voltage into the video cables, causing long-term problems such as poor picture quality or even serious equipment damage.

One final drawback to using existing light poles is that extension ladders do not work well when leaned against a round pole. This means that a lift or a tall folding ladder would be required, which is not usually part of the standard service equipment. An installation team may have a lift on site, particularly for newly constructed facilities, but this means the lift would again be required every time the cameras are serviced.

Installing cameras on their own independent poles will eliminate some of the concerns associated with using existing poles, but this also brings new concerns into play. If independent poles are used, the only high voltage present will be the wiring that is designed specifically for the camera. Induced voltage on the video cable will be less likely, and the risk of electrical shock will be less for the technician performing the installation.

Using independent poles, however, means that the poles must be installed into the ground, something that typical security installers are not frequently asked to perform. This means most likely an additional expense to have another contractor install the poles and the security technician to install the cameras. Because the poles are often located away from the buildings, the cabling must usually be installed by way of underground trenching. This would also be true when using existing poles. That will often mean digging up landscaped ground, sidewalks, parking lots, and anything else that may be in the path from the pole location to the building. Although this is not much of a concern for future service, it does create an additional expense during the initial installation and is just one more thing that could cause serious problems. There have been many cases where an unknowing technician has begun trenching for cabling without verifying what is underground first, resulting in cut phone lines, power cables, sewage lines, and more. This again is another risk that should be avoided if possible.

When using independent camera poles, it is also possible to select the wrong type of pole for an application or even the wrong location and mounting height. One camera location that I worked on demonstrates this point effectively. At this particular camera location, a camera with a zoom lens was installed in a large outdoor housing equipped with a heater and cooling fans. The camera housing assembly fully loaded weighed approximately 65 pounds. This assembly was installed on an older pan/tilt assembly, which weighed about 55 pounds. At some point in time the pan/tilt assembly ceased to work and had to be replaced. The entire unit was installed on a 13-inch-wide triangular antenna tower and was mounted at a height directly parallel to the fourth-floor windows of the nearby building. The tower was about 10 feet away from the building, which meant that it could not be reached from the building. Although the tower was mounted straight vertically, the ground in which it was installed was sloped at too steep an angle to be serviced from any lift available in the area. Another technician and I were called in to replace the defective pan/tilt unit with a new one that was identical in size. The only way to get up to the camera was to physically climb the four-story tower to the top, with me being the lucky one chosen to do so. Because the entire assembly was completely on top of the tower, the closest I could get was directly below the assembly.

To replace the pan/tilt unit, the first step was to remove the camera and housing assembly. Once the 65-pound unit was unbolted from the pan/tilt unit, it had to be lowered to the ground from above by a rope, which was quite a task in and of itself. Next, the pan/tilt unit had to be unbolted from the tower and lowered to the ground in the same manner. Doing all of this while holding on and being strapped to the antenna tower was time consuming and physically demanding just in the removal process. After nearly 1 hour on the tower, it was necessary to come down for a while just to regroup and get the blood flowing again.

After a brief rest, the new pan/tilt unit and the old camera assembly had to be installed on the tower. Due to the weight and limited number of arms, it was not possible to simply carry the pan/tilt up the tower and install it. It was also not physically possible to pull the pan/tilt and the camera assembly up with a rope in the same way as they had been lowered. The only viable alternative was to have the pan/tilt unit raised up to the top with a rope and pulley from the ground. While this may sound somewhat minor, the laws of physics became obvious quickly, such as every action causes an equal and opposite reaction. Every time the technician on the ground pulled the rope to raise the pan/tilt higher, the tower would sway toward the direction he was pulling. Although the tower was held with several guide wires, the fact was that it was only 13 inches wide and four stories tall. Although the amount of sway was probably relatively small, to someone strapped to the top it seemed like the tower was moving 4 or 5 feet back and forth. The closer the pan/tilt came to the top, the more the tower seemed to sway. Raising a 55-pound object four stories into the air was also difficult for the technician on the ground.

Once the pan/tilt unit reached the top, it had to be lifted overhead to its resting place and then bolted down. Just getting the pan/tilt unit into place took another 30 minutes. Next was the task of repeating the process with the camera assembly, which weighed 10 pounds more. Again, the sway of the tower seemed to be a lot, and now with both technicians physically worn, this step took an additional 40 minutes. Now with the pan/tilt unit replaced and the camera assembly placed back in the desired location, the equipment was connected and tested before I could begin the climb down from the tower. Beginning to untangle from the seated position with legs through the tower bars, I soon realized that my legs had fallen asleep in the hour and 20 minutes I had been on the tower. It took about 10 minutes more of standing strapped to the top to regain enough feeling and blood flow to climb down safely.

This entire process took two technicians about 3 hours each to replace a simple pan/tilt assembly. The sad irony was that the nearby building was only four stories tall, and the desired viewing area of the camera was away from the building. This entire camera assembly could have been installed at about the same height and with a better view of the desired subject if it had been installed on the roof of the building. In addition, the initial installation would have been much easier, cheaper, and faster. No tower or guidewires would have been required, no cable trenching would have been required, and cable routing would have been much simpler. The same pan/tilt replacement would have taken one technician about 15 to 20 minutes to accomplish if the camera location had been planned a little better during or before the installation phase.

Installation-Ready Dome Poles

One camera equipment manufacturer has developed a unique pole assembly that deserves to be mentioned here. Designed specifically with installation and future service in mind, this pole allows the camera to be installed onto the pole and adjusted and configured from the ground. The dome mounts to a mounting arm that is connected to a track that goes to the top of the pole. Once the camera is ready, a screw mechanism is turned with a standard drill that raises the dome and arm up on the track to the top of the pole. The screw connector is then locked into place inside the pole, preventing access by anyone who does not have the key.

Installation is relatively simple, and future service, cleaning, and repair will require no ladders, lifts, or scaffolding. With this type of configuration, maintenance and repairs can easily be done with only one technician and in much less time than with a traditional pole.

Roof Mounts

Another common mounting location for outdoor cameras is on the roof of the building being covered. This is another ideal location in many situations,

Figure 10.4 This Pelco parapet mount dome camera provides excellent accessibility. For service the pendant arm will swing in over the roof as pictured, giving the technician easy access without being in danger of falling. Also note that the end cap of the arm is removable for access to the cabling.

depending on the camera coverage desired and the height of the roof (see Figure 10.4).

Several types of camera mounts can be used for mounting a camera on a roof, and each is best when utilized for specific applications. The primary types of mounts are the standard roof mount, the parapet mount, the wall mount, and the pole mount. Usually a pole-mount configuration on the roof is much smaller than a stand-alone pole mount and is only used if a pole is available for a good mounting surface. The camera is usually then mounted at a height that will not require a ladder, and the pole is typically only 2 to 4 inches in diameter.

Wall mounts are used when it is fairly easy to reach over the parapet safely to install the mount (see Figure 10.5). This gets the camera out away from the building while still providing access to the camera. This type of mounting works

Figure 10.5   Pictured is a Kalatel dome camera with a pendant wall mount.

best when used with a camera enclosure that opens from the top for future service. A standard roof mount has a large base that is designed to sit flat on the roof. Typically these are best for buildings that have a very low parapet wall around the edge of the roof. Attached to the large base is usually an angled pole or mounting bracket that swings the camera out over the edge of the roof to obtain the best field of view. If the parapet wall around the roof of the building is too high, the camera-mounting bracket will not be able to easily clear it, and the camera may end up to close to the building or directly on top of it, limiting the viewing possibilities. If the roof is the type that is covered with a rubber membrane to prevent leaks, it may not be practical to use this type of mount either. If the anchors holding the mount must penetrate the rubber membrane to hold the assembly into place, it could void the warranty of the roof if the installation is not done properly and/or is not certified by the roofing company. This means that any future leaks might then become the responsibility of the installing company or the building owner. If this type of mount will be used on this type of roof, it is usually best to work directly with the roofing company to make sure that it is installed and anchored properly. A small pedestal is often installed on the roof in the desired location, covered with another rubber membrane, and sealed to prevent leaks. It is then possible to mount the base to this pedestal with good anchors. The anchors will penetrate the new rubber and the pedestal but will never penetrate the actual roof.

Most wall mounts do not allow for the camera assembly to be easily brought in to the roof of the building, meaning installation, maintenance, and repair must be done by reaching out to the camera location, particularly if a regular camera enclosure is used. This can be much more dangerous than a mount that swings the camera safely back to where the technician has safer access. For that reason, this is the least recommended style of mounting cameras on the roof. For dome-type cameras, wall mounts are made that do pivot up and back, bringing the camera either onto the roof or much closer to it. The pivoting type of wall mount will work well provided that the camera pivots in close enough for the technician to safely reach it. The disadvantage of this type of mount is that the camera must then be serviced upside down. In most cases, this is not a problem, but for some adjustments this may not always be practical.

Parapet mounts are designed to anchor directly to the parapet wall around the edge of the roof. They are typically mounted to the inside edge for ease of installation, but for some applications they can be installed on the outside edge over the wall of the roof. Parapet mounts are often used with dome enclosures and are designed so that the technician can swing the camera in onto the roof of the building to work on it.

These mounts work best on facilities where the parapet wall is high enough and sturdy enough to support the entire assembly. They are relatively easy to install and service, are not nearly as large and cumbersome as standard roof mounts, and usually allow the camera to be positioned farther away from the building than other mounts. If a camera is being mounted at the corner of the building, a parapet mount allows the mounting arm to be set into place at a 45-degree angle so a PTZ camera can see down both adjacent sides of the building. This can also be done with a standard roof mount, but often the camera is then not far enough out to adequately see both sides.

Service and Maintenance Access

Perhaps one of the most important things that you should remember from this chapter is the fact that future maintenance and repair access should be considered at the time of the system design and installation. Far too often the added labor, costs, risks, and inconvenience are not realized until the first time a camera requires service. The technician is often dispatched to repair a camera, only to discover that special tools and equipment are needed just to reach the camera. The technician's safety is unnecessarily jeopardized just because the installation location was not thoroughly thought out in advance. Added maintenance and repair costs are also not noticed until the service call shows that a lift is needed, additional technicians are needed, and extra time is needed simply because the camera location is not easily accessed. Mounting a pan/tilt unit upside down to the bottom of an eave may blend in well with the building architecture, but if and when that unit requires a replacement, it will be a service call nightmare. Pole mounting all the outside cameras may look the best for the environment, but getting to the camera may become much more difficult than it needs to be.

A few hours of planning in advance can easily save hours every year in future maintenance and repair costs, particularly on large systems. Two additional hours required to service one camera is not really much, but two hours per camera for 20 cameras means a week of lost labor and unnecessary labor costs.

CONCLUSION

If cameras are going to be installed outside a building, many factors should be considered in advance. Lighting, viewing subject, mounting heights, risk of vandalism, and future service requirements are essential factors that could decide whether the camera system is great or poor. If the wrong type of camera is selected or the wrong location is selected, the view from the camera can easily be less than desirable. If the future accessibility of the equipment and seasonal conditions are not considered, the camera location could cause long-term problems for the user.

This chapter does not cover every possible problem that may be encountered when installing outdoor cameras, but it is hoped that this information has made you more aware of how to best evaluate a particular facility or application. If you are currently designing a system or systems, it is hoped that you have gained enough insight from this chapter to help make the system or systems more user friendly—not just initially, but for years to come. For those who are responsible for existing camera systems, this chapter has provided you with enough information to evaluate current outdoor camera locations for possible shortcomings that can be corrected. Correcting these shortcomings now could save the company valuable dollars in the long run with maintenance and repair costs, or it could even save a technician from unnecessary risk of injury.

Tying It All Together

This chapter will take the information discussed in the previous chapters and use it for a practical sample application. This example will help show how the information from this book can be used in real-world applications.

The previous chapters have looked closely at particular fundamentals of planning and designing a camera system, but this chapter is configured a little differently. This chapter uses an imaginary facility and follows the process through from the initial evaluation to planning for long-term use and system growth. The final design for this imaginary facility is only one of many design concepts that could be used for a facility of that type and in no way should be viewed as the only correct design. As shown on each subject throughout this book, there are many variables, all of which are open to interpretation in different ways by the designers, technicians, and system users. The ultimate goal is not to show the only way to design a system, but to show the designer or end user how to determine what the best approach might be.

For ease of explanation and to keep this chapter brief, the imaginary facility will be an underground parking garage. This is not to say that a parking garage is the easiest facility to design a system for, merely that in this way the chapter can focus on just one aspect of facility design. If a full commercial facility were used, it would require multiple steps—from analyzing office areas to parking areas and shipping and receiving areas. Each of these would require review and planning in different ways and all would be tied together for the overall system design (see Figure 11.1).

The steps used when designing a system for a parking garage are the same steps that could be used when planning other areas of coverage. The entire process would be repeated or done simultaneously for all other unique areas of the facility. In a retail environment, for example, the designer would go through each phase for the cash office and/or cash lanes, the customer areas of the store, and warehouse areas. Although the same steps could be used for each area, the design

Figure 11.1 This parking garage uses multiple cameras to monitor for security and safety concerns.

concepts would be entirely different, because the coverage and recording needs would be different.

Before a system can be designed for any facility, it is important to find out a few crucial answers. All of the questions are fairly straightforward and should be easy to answer, but they also help make sure that the end users know exactly what they want.

PLANNING STAGE

All of the preliminary questions regarding the eventual system design will be done in the planning stage. This stage requires the most interactions between the designer and the end users to make sure that the system is designed to meet the end users' needs. If this phase is done completely and thoroughly, the end user will have no misconceptions of what the system will and will not do. The designer will also have no misconceptions about what the end user wants the system to do and how the system will be used.

Planning the system can be much easier if the designer first creates a few survey formats to get all of the answers needed to properly design the system. The information obtained in these surveys would then be compiled and reviewed with the end user to form a facility needs analysis. The length of the surveys will depend on the size of the facility and the number of unique areas of the building that must be covered. The parking garage, for example, has three unique areas to evaluate: (1) the entries and exits into the garage, (2) the general parking area for customers, and (3) the vestibule or lobby area where people will flow in and out of the attached facility. Camera coverage will most likely be different for all three areas because the needs are different.

User Survey

The parking garage is an underground garage for a retail building. The garage area is 190,000 square feet of parking for customers of the retail establishment. Employee parking is not located in the garage, so all vehicles should be customers only. No deliveries come through the garage area, so the coverage requirements should be fairly straightforward and easily definable. The parking garage is the property and responsibility of Motley's, an imaginary retail chain. The company has a full-time security staff responsible for all aspects of safety and security for the building. A security office has staff present at all times that the building is occupied.

The system designer should first create a series of questions that are fairly obvious to help with the eventual design. These questions should be asked of any of the responsible parties within the organization who may have some input. For example, the security director will definitely have some input on the system requirements, because his or her group will be responsible for using the system. The legal group, such as corporate counsel, will also have an input, because they are responsible for policies and procedures for the company. Regional and national security and legal managers may also have input to make sure that the facility planning is within the scope of the overall company objectives and image. Obviously, the survey should not be for just one person within the company unless that person is the only one responsible for all aspects of the facility, which is extremely rare.

Far too often not enough people are involved in the early stages of system planning. By not involving all decision makers and influential people, the system designer will not know for sure if the system is designed to meet the best needs of the customer. For example, if the only person involved in the planning stage is the security manager, the designer is assuming that the security manager knows all aspects of the company's concerns and legal requirements. The designer is also assuming that the security manager knows his or her job better than anyone else, will be there forever, and will be happy with the final results. This is not always a safe bet.

If the design is based solely on the input of the security manager and that person later leaves, who is to say that the replacement will have the same concepts and skill levels as the departing manager? If the designer receives input from all decision-making levels of the company, this will serve as a sort of check and balance for the designer and the company, making sure that the company is ultimately responsible for accepting the system design. This is the best way a designer can protect him or herself from error and omission problems that could turn up.

The following subsections represent the questions asked on the user survey and the answers that were obtained for use in the design of the parking garage. The number of questions that could be asked during this survey depends on the designer and his or her knowledge of the customer and facility type. For the purpose of this chapter, only a few key questions are used here.

What Does the User Hope to Achieve with This System?

Although this question is vague, it probably gets the most useful response from the end user. The answer to this question will usually cover the end user's perception of how the system will work and how the operators will interface with the equipment. If the answer is too vague, the designer may have to take the user's answer and ask more specific questions based on what is said.

The security manager for this facility gave the following answer to this question:

> I would like for the security officers to be able to watch the garage for any strange activity or problems that might require a response. If an incident occurs in the store, I want to be able to review the garage video to get a vehicle description and license plate number if the suspect comes in or goes out through the garage. I want to be able to track anyone or any vehicle through the garage in case of any suspicious activity. I also want to be able to look at any of the areas around the Emergency Call stations if they are activated.

The legal counsel gave this answer to the question:

This system should be set up to watch the garage to help limit the liability of the company from customers being attacked in the garage, slip and fall accidents, or other accidental injuries. This system should also interact with the emergency phones in the garage. Cameras should be set up to view the stairs and the elevator lobby as well.

Who Will Watch the Video Monitors and When?

This question will help the designer determine how much interaction will be involved between the operators and the control equipment. The security manager stated that the monitors would be observed at all times by a security officer, who is responsible for watching all of the cameras for the entire facility. For the retail store itself, there are 45 cameras to watch. The security station is monitored at all times that the store is opened to customers as well as times when just employees occupy the building. This means from Monday through Friday from 8:00 A.M. to 10:30 P.M., Saturdays from 6:30 A.M. to 11:30 P.M., and Sundays from 9:00 A.M. to 6:00 P.M. The legal counsel stated that the monitors should be watched any time that anyone is in the building.

Please Describe Other Security Measures for This Area

This section would list a few items that could be important for the overall system design and ask for a description of each. For this application, the security items listed would be lighting, Emergency Call stations, doors, vehicle gates, and public address systems. A section for other items should also be included, with plenty of room in each category for an adequate description.

For the example of the parking garage, the answers from the responding parties have been combined. Lighting for the garage consists of tungsten lamps mounted to the false ceiling throughout the garage. The entire garage is covered with fixtures spaced every 10 feet, and all fixtures are always lit. The company felt that it was important to have a bright and cheery parking garage so that the customers would feel upbeat and safe when they came in.

There is an Emergency Call station within 75 feet of any point in the garage. There are a total of 12 Emergency Call stations. Each station is mounted on a bright red building column. The station is a large bright yellow metal box with the word "EMERGENCY" written in black on each side (see Figure 11.2). On the front panel are the words "Push Button for Assistance." Above each station is a blue strobe light that is activated when the call button is pushed. The light stays on until the security officer resets it.

There are two large sliding-glass doors between the parking area and the elevator/escalator lobby. The security officers lock these doors at night once the building and parking garage are found to be clear. There are no vehicle gates at

Figure 11.2 This Emergency Call station by Louroe is tamper resistant and is often used for parking garages and parking lots.

the entrances or exits of the parking garage, nor are there any overhead doors to secure the area at night. When the building is unoccupied, it is possible for anyone to enter the garage area either in a vehicle or on foot.

As for the public address system, each Emergency Call station can be activated manually by the security officer to talk to someone in the vicinity. There is an intercom system that plays music and prerecorded shopping messages throughout the day. The security officer can also make emergency announcements on this system throughout the garage and the store. The security officer can also select just the garage area so announcements are not heard in the rest of the store.

Under the category of other, the security officer pointed out that there is a roving patrol of two security officers at all times for the entire store. At least one of the roving patrols must pass through the entire garage at least once every hour. After the store is closed, and every 2 hours throughout the night, a contract security company will send a patrol through the garage.

The legal counsel pointed out that there will also be signage throughout the garage stating that the company is not responsible for lost, stolen, or damaged vehicles and signs that basically state that the area is off limits outside of regular store hours. There are also signs that state that trespassing and loitering are not permitted and illegally parked vehicles will be towed at the owner's expense. Signs at the entrances and exits will state that there is no overnight parking and vehicles will be towed.

Additional Questions

These examples are only a few of the questions that could and should be asked during the planning stages of any system design. Questions should cover both facility generalities and specific concerns and needs. The more questions that are asked in the beginning, the easier and more detailed the system design will be for the designer. This also means a greater chance of 100 percent customer satisfaction with the system design and greater understanding by the end user of what the system will and will not do.

Site Survey

Another crucial task during the planning phase of the system design is the site survey. What is actually done during the site survey will depend on the status of the facility. In other words, if the site is not even built yet, then all of the site survey information could only be done with architectural and engineering drawings. If it is an existing facility that a camera system is being added to, there might not be any drawings available. That would mean that the site survey would just involve information physically gathered at the site. In an ideal situation, the designer would be able to do some survey and planning from the drawings and other portions from physically visiting the site.

Survey from Drawings

When performing a survey strictly from the drawings, the system designer must be familiar with reading schematics and blueprints. The designer must thoroughly understand architectural and engineering drawings and be knowledgeable in facility design as well as CCTV system design. If the designer is not familiar with facility design concepts, it would be easy to make an incorrect assumption that could affect the outcome of the system installation. For example, without thorough knowledge of architectural and engineering drawings, it might appear that cables could be routed through a section of a building, which in actuality is not possible. This would mean that the actual cables would need to be routed in a different direction, and the estimated amount of cable required would then be inaccurate.

When doing a survey or analysis from the drawings, the system designer should look at all possible camera locations and coverage patterns. Coverage concerns as noted from the user survey should be specifically addressed for proper coverage. In the parking garage, for example, it would be important to look at coverage requirements for the escalator/elevator lobby, the Emergency Call stations, and the entrances and exits for the garage. The other important area to look at is how the cables from the garage cameras will be routed to the control area within the store.

Survey on Site

When the survey is done strictly from a site visit, the designer must have a thorough understanding of what to look for and what to document at the site. It is highly recommended to take a large number of pictures or even videotape of the facility to go back to for any future questions that may arise. Every potential camera location should be looked at closely and even photographed to see how and where the camera would be mounted and how the cables could be routed to the location. Any special equipment or tools that might be needed should be documented to help in the design and estimating process to be done later. The designer will basically look at the same areas of concern as with the drawings, and it would be best if the designer has the capability to create drawings or have drawings created.

Survey from Drawings and Site Visit

If the designer is able to perform the survey from the drawings and from a site visit, then the survey will be much more thorough. Because the time on site will probably be more limited than the time with the drawings, the designer should take steps to ensure that the actual site time is the most efficient. The designer should thoroughly review the drawings before the site visit and make special note of any key areas that may not be clear from the drawings. Potential camera coverage areas should be noted so that they can be properly assessed and photographed while on site. Key factors, such as cable routing, should be determined from both the site survey and the drawings. For example, the desired cable route could be determined with the drawings, and then the site survey will verify whether the desired routing is actually possible.

When the survey is done from both drawings and a site visit, the designer should already have a basic concept in mind before going on the site visit. By working with the drawings, the designer should start to form an early design concept so that the site visit is most effective as a tool to verify thoughts, confirm that it can physically be done, and ensure that the planned coverage will be adequate.

Needs Analysis

The term *needs analysis* in this context may be slightly different from how you would typically use it. For the purposes of this book, the needs analysis is a document that covers what the needs of the end user will be for the camera system. This needs analysis is essentially a tool that the designer can use when discussing the system with the end user to portray how the system will be designed, what key functions and capabilities it will include, and how the system will work. The needs analysis should take the information from the user and site surveys and describe how the system will cover everything addressed in them. It should not cover specific equipment choices but should cover general design concepts in detail.

Parking Garage Needs Analysis

The following statement would be a typical excerpt I might use in a needs analysis for the garage system:

> This document is a needs analysis and early design concept for the Dewey Cheatum and Howe (DC&H) parking garage. It is meant to serve as an initial concept document to help further clarify any system design criteria before the selection of the system-specific equipment.
>
> This camera system consists of 24 cameras within the underground parking garage. One camera with pan/tilt and zoom (PTZ) capabilities will be located at the primary entrance to the garage. One camera with PTZ capabilities will be located in the main pedestrian vestibule for coverage of the escalator and elevators. To provide coverage of the Emergency Call stations, one fixed camera will be used for each station. If a button is pressed to call for assistance on any station, the associated camera will respond to that as an alarm and will pop up on a separate monitor dedicated to cameras with alarm activity. In addition, the recording rate for that camera will be increased until a security officer resets the alarm. Should multiple stations be activated, all cameras with alarm activity will be displayed on the

alarm monitor in sequence until each is reset. The recording rate for each will also be increased.

Two additional fixed cameras will be used in the pedestrian vestibule for coverage of the escalator, elevators, and exit doors. The remaining eight cameras will be placed in the parking area to cover the secondary exits and primary vehicle paths.

All cameras will be color cameras in either domes or enclosures to protect them from vandalism and dirt from vehicle exhaust. Varifocal (manual zoom) lenses will be used on all fixed cameras to ensure that each camera has the desired coverage pattern. All camera cables will be routed to a common conduit located next to the support column K7. This conduit will run through the ceiling into the security office located directly above. This will require a 4-inch-diameter hole to be core drilled through the floor of the security office. The conduit will route to an interface cabinet, which will allow the cameras to be connected to the camera control equipment. The control equipment will be mounted in a fixed 45-inch-tall cabinet with a locking door and ventilation fan. Viewing monitors will be installed in a sloped security workstation with secondary monitors located in the security manager's office.

Control equipment will have several options, which will be determined based on the budgetary restrictions and requirements of the project. An analog and a digital option will be provided with detailed projected long-term costs for each, as well as the initial costs. The analog option will include two duplex multiplexers, two video recorders capable of the maximum recording time and the highest frame rate for the facility, and a controller for the PTZ cameras. The controller may be part of the multiplexer depending on the type of multiplexer, and cameras chosen.

The digital option will consist of either two multiplexers and two digital video recorders or two multiplexing digital video recorders, depending on the manufacturer selected. DC&H will be shown the difference between the two types and will be instrumental in the final choice. The digital system will also be equipped with some type of video archiving to be determined based on the digital system selected.

After the equipment selection process for the specific equipment locations, all options available will be presented to DC&H to help determine the final system design. The needs analysis will also be reviewed at that time to determine if any additional requirements or concerns should be addressed.

DESIGN STAGE

Once the needs analysis has been accepted by the end user, the designer must start the formal system design. Although much of the actual design has already been determined, the formal design has not. Based on the needs analysis document and any feedback from the end user, the designer will work on the minute details of the design to be sure that nothing is overlooked. This will include determining the exact camera locations and cable runs, specifying all mounting hardware and equipment required, and preparing for equipment selection and cost analysis.

Control Equipment Location

In the example of the parking garage, the customer predetermined the control equipment location. This is often the case, because the customer already knows what functions will be performed in each section of the building. Occasionally the designer will have some input regarding the control location, however, and should evaluate the requirements accordingly. If the location were not selected by the end user in advance, the designer should find out in the survey phase who will be monitoring the equipment, if multiple locations are required, and if multiple access levels will be required.

Camera Layout

The needs analysis summarized the general coverage of the cameras, but the designer must now look at specific locations and coverage patterns. This would be the time that the designer looks at the specific aspects associated with each individual camera location, such as parameters required for the camera, mounting brackets required, type of housing to be used, and what additional features or functions are needed.

The designer should determine the exact coverage areas and layout patterns for the cameras and whether features such as digital signal processing (DSP) and backlight compensation will be required. The designer should also look at whether heating and cooling elements will be needed or if the garage will stay within the operating temperature range of most cameras.

Alarm Devices

As was suggested in the user survey, some alarm triggers will be required for this system. In this case, most of the alarm triggers have already been determined, so the designer will just have to configure how the triggers will be interfaced with the camera system.

The user has stated an interest in having the cameras react to any activation of the Emergency Call stations, and the designer has already planned for this based on the comments in the needs analysis. The only effect that this has on the camera system is that additional wiring will be required, either from the Emergency Call stations or from the Emergency Call control equipment, depending on the type of systems chosen and how they interface with each other. This area would require further research by the designer to see which ways are possible and which way is the most efficient. For the purpose of this example, the system in place requires that an additional cable be run from each Emergency Call station to an alarm-input zone on the camera control equipment. The Emergency Call station requirements state that only 18-gauge or larger cable is used and that it must be stranded and shielded. For 18-gauge cable, the maximum wire run length cannot exceed 1,500 feet.

System Cabling

At this stage in the system design, the designer should know approximately how far it is from each camera to the control equipment in the security office. The design shows that the longest camera cable run is 1,265 feet and the shortest is 185 feet. The designer has a few choices regarding the types of cable that could be used and many different ways to configure the system.

It would be possible, for example, for the designer to use RG/59 cable for the video signal. Most of the camera runs are less than 750 feet, and they would have no problems with providing an adequate video signal. A few of the cable runs are over 750 feet, though, so they would require either video amplifiers, some type of signal boost as is available with some multiplexer manufacturers, or a different type of cable. At 1,265 feet, RG/6 cable could easily be used. The designer in this case, however, has decided that the same type of cabling will be used for the entire system and that the added expense of the RG/6 and the additional labor could be better utilized elsewhere.

Fiber-optic cable could also be used and would provide an excellent video signal. The designer has looked at this possibility and has come up with a few areas of concern with using fiber. The technicians who will be installing the system, for example, have limited experience with fiber, and the service technicians have even less experience. Also, using fiber would mean additional equipment for transmitting and receiving the video signal, additional expense for the higher price of fiber, and additional labor to install the fiber ends.

Transmission over twisted-pair cables was another possibility for the garage cameras. The cable is cheaper than the RG/6 and the fiber, and the technicians are experienced in working with it. Additional equipment would be required to transmit and receive the signal across the twisted-pair cabling, but labor would be lower and the cost difference of the cable more than makes up for the additional equipment expense.

The designer found that for the nature of the facility, conduit would not be required for the entire cable route with most cameras. Because the garage has a false ceiling, most of the cable could be installed above the ceiling with limited risk of tampering. At any place in which the wire path would be exposed, the designer planned for conduit, junction boxes, and flexible conduit to fully protect the cables. This also gave the casual observer the impression that the system was well protected to discourage any tampering or investigating by someone passing through.

End-User Review

At this stage in the system design, as with any other milestone, it is important that the end user be given the opportunity to review progress and make any possible changes. By allowing the user to review the plan at each step of the way, neither the designer nor the user will have any major surprises when the project is too far along to make minor adjustments. At this review for the example site, the user fully agrees with the decisions the designer has made thus far. The users may also feel reassured that they are getting a design that is custom tailored to fit their exact needs, and by being a part of the decision process they will feel more in control of the entire project.

EQUIPMENT SELECTION STAGE

Now that the designer has a complete system design plan, it is time to start the equipment selection process. As the design has progressed through each phase, the types of equipment that will be used have been narrowed down so that the number of possibilities will not overwhelm the designer. For example, it is now known that the system will be a color system with sufficient lighting for virtually all color cameras. The lighting is of the type that will provide a sharp, clear, high-quality color picture without any major surprises. It is also known that 24 cameras will be used, which will help when deciding on the quantity and models of the control equipment. Only two of the cameras will have PTZ capabilities; the others will all be fixed cameras with Varifocal lenses. The lighting is consistent throughout the parking garage, so only two camera models must be selected.

Choosing the Cameras

When choosing the cameras, the designer must evaluate what specifications are required for each camera type and location. In the parking garage example, there are only two camera types. The first types of camera that the designer will look at are the two PTZ cameras. One of these units will cover the main entrance into the garage. It must be capable of capturing the license plate information from a vehicle exiting or entering the facility. Because vehicles may enter with their headlights on, the designer must plan for this possibility. That means that the camera should have DSP and/or backlight compensation to view the vehicle's license plate effectively.

Because the designer is unsure of the possible vehicle speeds when entering and exiting, the camera should be able to pan and tilt at a fairly quick rate. Also, to make sure that no vehicle is missed, the designer may want to select a unit that has alarm triggers at the camera location to automatically return the camera to a preset position. A photobeam across the main entrance could cause the camera to automatically return to this proper position quickly, catching the vehicle before it is too late. To make it a little more accurate, one set of photobeams could be used for the entrance lane and another for the exit lane, each with its own camera preset position to ensure the best possible image.

Next, the designer would look at the requirements for the fixed cameras in the garage. As already mentioned, the cameras would be color and would have a Varifocal lens. This would narrow down the field of potential cameras, because many of the small dome and bullet cameras do not have Varifocal capabilities. As with the PTZ cameras, headlights from the vehicles could cause potential backlight problems, so either DSP or backlight compensation may be important. The designer should narrow the field of choices, and then compare the specifications and pricing of each to help determine which option provides the best solution for the end user.

After evaluating the choices and narrowing them down, the designer has decided on high-speed domes for the PTZ cameras and smaller fixed domes for the fixed cameras. Part of the decision was based on the durability and reliability of the domes, and part was based on the esthetics of the domes. The cameras selected would easily fit in with the bright atmosphere of the garage and would look more rugged and less industrial than cameras in enclosures mounted on the walls.

Choosing the Control Equipment

As was stated in the needs analysis, the designer will provide two options to the end user as far as the control equipment. The designer has elected to do this for several reasons:

1. *Places the final decision in the user's hands.* The number of choices is limited, so it should not be a major decision that takes forever.

2. *Provides two different price ranges for the user.* This allows the user to select a system that fits the budgetary requirements while still meeting the security needs.

3. *Shows the customer the primary differences between digital and analog.* It will show the initial cost difference, the long-term cost difference, and the capabilities difference if the designer has packaged everything properly. In addition to providing a cost comparison between the two system types, the designer should show a capabilities comparison between the two to indicate what benefits can be gained from either selection.

The designer in this example has presented two system types with a common tie. With the analog system, the designer selected a multiplexer with built-in alarm capabilities, PTZ camera controls, video motion detection, and easy expansion with multiple multiplexers linked together. The system will require two multiplexers capable of handling 16 cameras each, leaving expansion room for 8 more cameras. Each multiplexer is connected to a 24-hour real-time recorder with a record rate of 20 frames per second. The multiplexer is capable of having more than one controller, so one can be located in the security office and one can be located in the security manager's office.

The digital design is almost identical to the analog design, with one key difference. The digital system uses the same multiplexer, but instead of a real-time recorder, it uses a digital recorder, which is directly interchangeable. The digital recorder is capable of up to 60 frames per second, and the record rate and file size for each camera is individually selectable. The digital images are recorded to a hard drive, and the system has a SCSI port for connection of an archiving device. The designer has chosen a DDS3 tape carousel for the archiving device, which is capable of holding six tapes of 13 gigabytes each.

Choosing the Alarm Devices

During the design phase, the designer should have evaluated where alarm triggers could be used to enhance the recording and tracking capabilities of the camera system. In the parking garage example, a few triggers were apparent that could do this effectively. The Emergency Call stations, for example, were already in place and provided an auxiliary relay specifically for camera systems. Two sliding doors into the elevator area meant that door contacts could also be used to enhance the recording capabilities and/or move the PTZ camera to preset positions. At the main entrance and exit into the garage, two sets of photobeams could be used to ensure that the PTZ camera always caught a car entering or exiting the garage. Because the rest of the garage was wide open, motion detectors would do little more than use the video motion detection capabilities already built into the multiplexer, so nothing would be gained by adding them.

Choosing the Cabling and Equipment

During the design phase, the designer decided to use twisted-pair cabling for the video signal from each camera. This meant that the designer had only to decide on which manufacturer to use for the transmitters and receivers required at both ends of the cable. At the control equipment end, it was more practical to use panels capable of receiving and converting signals from multiple cameras. Two primary manufacturers made units called *hubs*, which would handle 16 cameras each and required minimal rack space in the equipment rack. This meant that the designer needed to use two hubs, one for each multiplexer. For the camera side of the cables, a total of 24 transmitters would be required, eight of which had to be capable of transmitting the video signal over 1,000 feet. Both manufacturers made equipment that was capable of all of the requirements and had similar specifications other than price. In this case, price would be the deciding factor for the system designer.

At this stage of the equipment selection, the designer must evaluate everything to make sure that all necessary items are included. Everything from mounting anchors to conduit fittings should be included to make sure that nothing is missed in the pricing. Once the complete equipment list and pricing are developed, the labor requirement must be determined. This will require someone experienced in project management and possibly in installation to fully evaluate. The labor requirement as far as hours needed is outside the scope of this book and will not be included. For the purpose of the parking garage example and comparison of the digital and analog systems, an additional 10 hours of labor was required for the digital system, mainly for programming and end-user training.

End-User Review

By now the designer has fully developed the system design and any options that the end user might be able to choose. All of the information would then be organized into either a formal proposal or a presentation to explain the system and the differences to the user. Pricing, capabilities, expandability, mean time between failure, and long-term costs should all be compared and put into easy-to-understand documents for the end user. The designer and users should go over the entire system design to decide what changes (if any) should be made, which choice is more cost effective, and which choice is better for the company in the long run.

After a few lengthy discussions, the user in the garage example has decided to go with the analog system proposed by the designer. The deciding factor in the end-user's decision was that the analog system was cheaper than the digital system by a few thousand dollars. The other deciding factor was that the designer designed the system so that the end user could easily upgrade to the digital system when the budget allowed by simply changing the recorders. Because the end user already had an idea of what he wanted to spend for the system, only the analog system met the immediate budget. Knowing the upgrade costs at the time of installation meant that the user could plan for the cost requirements for next year. The designer also provided the end user with the cost justification for the later upgrade by including the mean time between failure and long-term maintenance costs for both systems during the initial design phase.

SYSTEM INSTALLATION

Proper installation of the camera system is a key factor in the ultimate performance of the system. Although the main focus of this book is to understand the components and design principles of a camera system, it is also important to mention some aspects of the installation to avoid long-term problems.

Planning the Installation

The end user of the system should discuss with the installation company the schedule and estimated time frame of the installation. For small projects, this may not be complex, but for new building construction or large systems, this point can be important to ensure that there are no disruptions. The basic work-flow of most installations is planning, cabling, device installation, equipment connection, setup, programming, testing, training, and final delivery. At each stage of this procedure, the user should check to make sure that everything is going as planned and that there are no major problems. The installation manager should also check at each phase to make sure there are no problems. This way, if anything comes up that will require changes to the system or additional work and funding, it will be known early enough that a decision can be made without creating hard feelings between the two parties. Communication can often help both parties in the long run to ensure that everybody gets what he or she wants.

Managing the Installation

For some large system installations, it may be important to have a person desig-nated as a project manager. This person would be responsible for overseeing the project from start to finish and usually would act as the liaison between the user and the installation company. The project manager should be familiar with this type of project and with the user's requirements. The project manager should also be familiar with the system design so that each phase of the installation can be accurately evaluated.

If the project is not large enough to justify a project manager, the represen-tative from the end-user's company and the representative from the installation company should work closely through all phases of the installation. Basically, the two people will work together to perform the same functions that the project manager would normally do, except on a smaller scale.

Reviewing and Accepting the Installation

Once the system has been installed completely, it should be fully evaluated to make sure it is exactly what both parties said it would be. Every item, including the cabling, should be reviewed to make sure it was installed exactly according to the design. All cabling and devices should also be checked to verify that they are installed in a neat and professional manner and in compliance with any applicable codes. A few of the codes and regulations that may affect a camera system installation are the National Electrical Code, International Building Code, and Americans with Disabilities Act (ADA) requirements. You should check with local experts or authorities in the area to help determine what local, national, or international requirements may apply. A reputable installation company will be fully aware of any requirements and should be happy to go over them with the end user. Compliance with any and all requirements should be part of any final acceptance checklist.

During the final inspection, the end user or the installation manager may come across a few items that should be corrected. All of these items should be put together in what is known as a punch list so they can be corrected and easily reviewed on completion. The final inspection should be thorough and detailed for the protection of both the end user and the installation company. Far too often, the system is simply turned over to the user, who begins using it. After a period of time, if something is brought up that could be a problem, there is no way of knowing who is responsible for correcting it. Most often the end user is under the assumption that the installation company did something during the installation, and the installation company is under the impression that it is something the end user or another contractor did after it left. This can only lead to hard feelings and a damaged relationship between the two parties. If a thorough inspection is done at the completion of the installation, there will usually not be any disputes.

In fact, I highly recommend that the installation company take detailed digital pictures of every device and screen shot on the system if it is allowed. Images with a time and date stamp should be taken, showing how the device looked when it was installed and how the image looked on the monitor. A complete copy of these pictures should also be provided to the end user on completion of the project, should he or she need it in the future. This has helped end users on occasion when another contractor damaged a camera after the installation.

In one instance, the picture of a camera location clearly showed the camera undamaged on a certain date. Two days later, the camera was damaged, obviously from being hit with a piece of heavy machinery. The contractor in question swore that it was like that before it was in. Videotape from the system showed that this contractor was the only contractor in the building at the time. Unfortunately, the building was still in construction, so only a test tape was being used in the recorder and had stopped recording before the camera was hit, but the picture with a time and date stamp and the videotape of the contractor entering the building was enough for one of the contractor's employees to admit that he had hit the camera (and the wall) with a scissors lift and failed to tell anyone.

System Documentation

Particularly with large camera systems, documentation of the installation can be important. Drawings and diagrams, pictures, programming sheets, and equipment lists can be useful to the end user and the technician responsible for long-term service and maintenance.

As-built drawings are the drawings made during the installation that show exactly how the system was installed. These drawings may be the same as the design drawings, but in many cases they are different because of additions and changes made during the installation. As-built drawings should also include equipment locations and wire numbers in some manner to make it easy to cross-reference with the other documentation. End users should note that if as-built drawings are desired or required, the installation company should be made aware of that fact before starting the installation.

USING THE SYSTEM

Once it has been determined that the system is completely installed according to the initial design, it is time for the system operators and security staff to learn how to use the system. System training should include much more than a few simple instructions given in a 15-minute session. Training should be an in-depth part of any system, with full training materials and shortcut keys for the operator. A tabbed index of possible actions should be kept somewhere close to the main system control equipment so that an operator can refer to it quickly if needed.

Technical Training

Detailed technical training should be provided to anyone who might be required to make programming changes to the control equipment. This training should cover operator training to ensure that the person is familiar with how to use the system. Once all potential operators are comfortable with the operation of the system, they can be trained how to make essential changes, such as changing passwords, changing video motion detection coverage patterns, and changing alarm responses. They should know how to change the time and date, how to change any camera description text on the screen, and how to set different user levels if the equipment provides that capability.

Depending on the size of the installation and the amount of access the user will require, the technical training could be anywhere from a minimum of one or two hours to a couple of days. The user would be best advised to ask as many questions as possible and practice with the equipment when feasible.

User Training

User training will not be as in depth as the technical training but may take just as long. The actual system users will need to be much more familiar with every aspect of the system than the manager who only uses it occasionally. Training may need to be done in multiple groups, depending on the capabilities of the system and the multiple user access levels, if there are any. Either the installation company or the end user should create or obtain small guides for the system operators to refer to in case they forget how to perform a system function. Periodic refresher training should also be done and could be assigned as part of the preventive maintenance schedule.

PERIODIC REVIEW AND ANALYSIS

Once the camera system has been operational and in use for a while, it should be reviewed and compared with the initial concept to see if anything has changed. The user's needs or the coverage areas needed can often change for any number of reasons. If the system is reviewed and analyzed periodically, it can be updated and improved on without becoming obsolete and useless.

Evaluating Changing Needs

Some areas that may require evaluation could be the result in changes to traffic patterns or use of the facility. For example, a system put in and maintained by a building owner may meet the needs of a tenant, but if the tenant changes, the system may not be effective for the new building occupant.

Other factors such as changing neighborhood crime rates, increased number of employees, increased risk, and severity of threats can all have an effect on the effectiveness of the system. For example, when this book was started, the risk of a terrorist attack in a shopping mall in the United States was very low. Now, in light of the attacks on the World Trade Center and the Pentagon on September 11, 2001, a terrorist attack on a retail facility must be considered a viable threat at least on some level, particularly on large facilities in key areas of the country. Structures such as the Mall of America in Minnesota, Disney World in Florida, or other large compounds that draw a crowd must be viewed as potential targets. This could eventually change how camera systems are used and to what extent cameras are located.

A few years ago the idea of Big Brother meant constant government monitoring and was a cry often heard by those installing camera systems. Now, however, cameras are almost welcomed by employees, customers, and all. It is all a matter of perception, and currently the general perception is that the more we are protected the better off we are, even if it means giving up a little freedom in the process.

System Expansion

Another reason for periodic system evaluation is the expansion of a facility. A school, for example, could add another wing, which would require more cameras. This would mean more recorders, more control equipment, and more for the system operator to keep track of. The user should be aware in advance of what the expansion capabilities of the system are and what will be required to upgrade the system should it exceed its maximum size.

A system with 30 cameras and a matrix system capable of handling 32 cameras, for example, can only add 2 more cameras. If three additional cameras were required, either an additional matrix switcher or a larger one would be required. That makes the third camera installation much more cost prohibitive, and the user is then likely to avoid that addition if at all possible. If advanced planning had been done in the design phase, it may have been noticed that the facility was likely to expand in the future and more than two additional cameras would probably be required eventually. A different matrix switcher may have been selected initially even if it meant starting with 28 cameras instead of 30 to help balance out the increased cost difference.

Common Terms Used in CCTV

Aberration: A term from optics that refers to anything affecting the fidelity of the image in regards to the original scene.

AC: Alternating current.

AC/DC: Alternating current/direct current.

Activity detection: Refers to a method built into some multiplexers for detecting movement within the camera's field of view (connected to the multiplexer), which is then used to improve the camera recording update rate.

A/D (or AD): Refers to analog-to-digital conversion.

ADC: Analog-to-digital conversion. This is usually the first stage of an electronic device that processes signals into digital format. The signal can be video, audio, control output, and similar.

AGC: Automatic gain control. A section in an electronic circuit that has feedback and regulates a certain voltage level to fall within predetermined margins.

ALC: Automatic light control. A part of the electronics of an automatic-iris lens that has a function similar to backlight compensation in photography.

Aliasing: An occurrence of sampled data interference. This can occur in CCD image projection of high spatial frequencies and is also known as *moiré patterning*. It can be minimized by a technique known as *optical low pass filtering*.

Alphanumeric video generator (also text inserter): A device for providing additional information, normally superimposed on the picture being displayed; this can range from one or two characters to full-screen alphanumeric text. Such generators use the incoming video signal sync pulses as a reference point for the text insertion position, which means if the video signal is of poor quality, the text stability will also be of poor quality.

Amplitude: The maximum value of a varying waveform.

Analog signal: Representation of data by continuously varying quantities. An analog electrical signal has a different value of volts or amperes for electrical representation of the original excitement (sound, light) within the dynamic range of the system.

ANSI: American National Standards Institute.

Anti aliasing: A procedure employed to eliminate or reduce (by smoothing and filtering) the aliasing effects.

Aperture: The opening of a lens that controls the amount of light reaching the surface of the pickup device. The size of the aperture is controlled by the iris adjustment. By increasing the f-stop number (e.g., f-1.4, f-1.8, f-2.8), less light is permitted to pass to the pickup device.

Apostilb: A photometric unit for measuring luminance where, instead of candelas, lumens are used to measure the luminous flux of a source.

Archive: Long-term offline storage. In digital systems, pictures are generally archived onto some form of hard disk, magnetic tape, floppy disk, or DAT cartridge.

Artifacts: Undesirable elements or defects in a video picture. These may occur naturally in the video process and must be eliminated in order to achieve a high-quality picture. The most common are cross-color and cross-luminance.

ASCII: American Standard Code for Information Interchange. A 128-character set that includes the uppercase and lowercase English alphabet, numerals, special symbols, and 32 control codes. A seven-bit binary number represents each character. Therefore, one ASCII-encoded character can be stored in one byte of computer memory.

Aspect ratio: This is the ratio between the width and height of a television or cinema picture display. The present aspect ratio of the television screen is 4:3, which means four units wide by three units high. Such aspect ratio was elected in the early days of television, when most movies were of the same format. The new high-definition television format proposes a 16:9 aspect ratio.

Aspherical lens: A lens that has an aspherical surface. It is harder and more expensive to manufacture, but it offers certain advantages over a normal spherical lens.

Astigmatism: The uneven foreground and background blur that is in an image.

Asynchronous: Lacking synchronization. In video, a signal is asynchronous when its timing differs from that of the system reference signal. A foreign video signal is asynchronous before a local frame synchronizer treats it.

ATM: Asynchronous transfer mode. A transporting and switching method in which information does not occur periodically with respect to some reference, such as a frame pattern.

ATSC: Advanced Television System Committee (think of it as a modern NTSC). An American committee involved in creating the high-definition television standards.

Attenuation: The decrease in magnitude of a wave, or a signal, as it travels through a medium or an electric system. It is measured in decibels (db).

Attenuator: A circuit that provides reduction of the amplitude of an electrical signal without introducing appreciable phase or frequency distortion.

Auto-iris (AI): An automatic method of varying the size of a lens aperture in response to changes in scene illumination.

AWG: American wire gauge. A wire diameter specification based on the American standard. The smaller the AWG number, the larger the wire diameter.

Back-focus: A procedure of adjusting the physical position of the CCD-chip/lens to achieve the correct focus for all focal-length settings (especially critical with zoom lenses).

Back porch: 1. The portion of a video signal that occurs during blanking from the end of horizontal sync to the beginning of active video. 2. The blanking signal portion that lies between the trailing edge of a horizontal sync pulse and the trailing edge of the corresponding blanking pulse. Color burst is located on the back porch.

Balanced signal: In CCTV this refers to a type of video signal transmission through a twisted-pair cable. It is called balanced because the signal travels through both wires, thus being equally exposed to the external interference, so by the time the signal gets to the receiving end, the noise will be canceled out at the input of a differential buffer stage.

Balun: This device is used to match or transform an unbalanced coaxial cable to a balanced twisted-pair system.

Bandwidth: The complete range of frequencies over which a circuit or electronic system can function with minimal signal loss, usually measured to the point of less than 3 db. In PAL systems, the bandwidth limits the maximum visible frequency to 5.5 MHz; in NTSC, to 4.2 MHz. The ITU 601 luminance channel sampling frequency of 13.5 MHz was chosen to permit faithful digital representation of the PAL and NTSC luminance bandwidths without aliasing.

Baseband: The frequency band occupied by the aggregate of the signals used to modulate a carrier before they combine with the carrier in the modulation process. In CCTV most signals are in the baseband.

Baud: Data rate, named after Maurice Emile Baud, which generally is equal to one bit per second (bit/s). Baud is equivalent to bits per second in cases where each signal event represents exactly one bit. Typically, the baud settings of two devices must match if the devices are to communicate with one another.

BER: Bit error rate. The ratio of received bits that are in error relative to the total number of bits received; used as a measure of noise-induced distortion in a digital bit stream. BER is expressed as a power of 10. For example, a one-bit error in 1 million bits is a BER of 10^{-6}.

Betamax: Sony's domestic video recording format, a competitor of VHS.

Bias: Current or voltage applied to a circuit to set a reference operating level for proper circuit performance, such as the high-frequency bias current applied to an audio recording head to improve linear performance and reduce distortion.

Binary: A base 2 numbering system using the two digits 0 and 1 (as opposed to ten digits [0–9] in the decimal system). In computer systems, the binary digits are represented by two different voltages or currents, one corresponding to 0 and another corresponding to 1. All computer programs are executed in binary form.

Bipolar: A signal containing both positive-going and negative-going amplitude. May also contain a zero amplitude state.

B-ISDN: Broadband integrated services digital network. An improved ISDN, composed of an intelligent combination of more ISDN channels into one that can transmit more data per second.

Bit: A contraction of binary digit. Elementary digital information that can only be 0 or 1. The smallest part of information in a binary notation system. A bit is a single 1 or 0. A group of bits, such as 8 bits or 16 bits, compose a byte. The number of bits in a byte depends on the processing system being used. Typical byte sizes are 8, 16, and 32.

Bitmap (BMP): A pixel-by-pixel description of an image. Each pixel is a separate element. Also a computer file format.

Bit rate: Bps = bytes per second, bps = bits per second. The digital equivalent of bandwidth, bit rate is measured in bits per second. It is used to express the rate at which the compressed bitstream is transmitted. The higher the bit rate, the more information that can be carried.

Blackburst (color-black): A composite color video signal. The signal has composite sync, reference burst, and a black video signal, which is usually at a level of 7.5 IRE (50 mV) above the blanking level.

Black level: A part of the video signal, close to the sync level, but slightly above it (usually 20 mV–50 mV) in order to be distinguished from the blanking level. It electronically represents the black part of an image, whereas the white part is equivalent to 0.7 V from the sync level.

Blanking level: The beginning of the video signal information in the signal's waveform. It resides at a reference point taken as 0 V, which is 300 mV above the lowest part of the sync pulses. Also known as *pedestal,* the level of a video signal that separates the range that contains the picture information from the range that contains the synchronizing information.

Blooming: The defocusing of regions of a picture where brightness is excessive.

BNC: BNC stands for Bayonet-Neil-Concelman connector, and it is the most popular connector in CCTV and broadcast TV for transmitting a basic bandwidth video signal over a coaxial cable.

B-picture: Bidirectionally predictive coded picture; an MPEG term for a picture that is coded using motion-compensated prediction from a past and/or future reference picture.

Braid: A group of textile or metallic filaments interwoven to form a tubular structure that may be applied over one or more wires or flattened to form a strap.

Brightness: In NTSC and PAL video signals, the brightness information at any particular instant in a picture is conveyed by the corresponding instantaneous DC level of active video. Brightness control is an adjustment of setup (black level, black reference).

Burst (color burst): Seven to nine cycles (NTSC) or ten cycles (PAL) of subcarrier placed near the end of horizontal blanking to serve as the phase (color) reference for the modulated color subcarrier. Burst serves as the reference for establishing the picture color.

Bus: In computer architecture, a path over which information travels internally among various components of a system and is available to each of the components.

Byte: A digital word made of eight bits (zeros and ones).

Cable equalization: The process of altering the frequency response of a video amplifier to compensate for high-frequency losses in coaxial cable.

CAD: Computer-aided design. This usually refers to a system design that uses specialized computer software.

Candela (cd): A unit for measuring luminous intensity. One candela is approximately equal to the amount of light energy generated by an ordinary candle. Since 1948, a more precise definition of a candela has become "the luminous intensity of a black body heated up to a temperature at which platinum converges from a liquid state to a solid."

CATV: Community antenna television.

C-band: A range of microwave frequencies, 3.7–4.2 GHz, commonly used for satellite communications.

CCD: Charge-coupled device. The New-Age imaging device, replacing the old tubes. When first invented in the 1970s, it was initially intended to be used as a memory device. Most often used in cameras, but also in telecine, fax machines, scanners, and so on.

CCD aperture: The proportion of the total area of a CCD chip that is photosensitive.

CCIR: Comité Consultatif Internationale de Radiocommuniqué or, in English, Consultative Committee for International Radio, which is the European standardization body that has set the standards for television in Europe. It was initially monochrome; therefore, today the term CCIR is usually used to refer to monochrome cameras that are used in PAL countries.

CCIR 601: An international standard (renamed ITU 601) for component digital television that was derived from the SMPTE RP1 25 and EBU 3246E standards. ITU 601 defines the sampling systems, matrix values, and filter characteristics for Y, Cr, Cb, and RGB component digital television. It establishes a 4:2:2 sampling scheme at 13.5 MHz for the luminance channel and 6.75 MHz for the chrominance channels with eight-bit digitizing for each channel. These sample frequencies were chosen because they work for both 525-line 60 Hz and 625-line 50 Hz component video systems. The term 4:2:2 refers to the ratio of the number of luminance channel samples to the number of chrominance channel samples; for every four luminance samples, the chrominance channels are each sampled twice. The Dl digital videotape format conforms to ITU 601.

CCIR 656: The international standard (renamed ITU 656) defining the electrical and mechanical interfaces for digital television equipment operating according to the ITU 601 standard. ITU 656 defines both the parallel and serial connector pinouts, as well as the blanking, sync, and multiplexing schemes used in both parallel and serial interfaces.

CCTV: Closed-circuit television. A television system intended for only a limited number of viewers, as opposed to broadcast TV.

CCTV camera: A unit containing an imaging device that produces a video signal in the basic bandwidth.

CCTV installation: A CCTV system, or an associated group of systems, together with all necessary hardware, auxiliary lighting, and so on, located at the protected site.

CCTV system: An arrangement comprising a camera and lens with all ancillary equipment required for surveillance of a specific protected area.

CCVE: Stands for closed-circuit video equipment. An alternative acronym for CCTV.

CD: Compact disk. A standard of media as proposed by Philips and Sony, where music and data are stored in digital format.

CD-ROM: Compact disk read-only memory. The total capacity of a CD-ROM when storing data is 640 MB.

CDS: Correlated double sampling. A technique used in the design of some CCD cameras that reduces the video signal noise generated by the chip.

CFA: Color filter array. A set of optical pixel filters used in single-chip color CCD cameras to produce the color components of a video signal.

Chip: An integrated circuit in which all the components are microfabricated on a tiny piece of silicon or similar material.

Chroma crawl: An artifact of encoded video, also known as *dot crawl* or *cross-luminance*. Occurs in the video picture around the edges of highly saturated colors as a continuous series of crawling dots and is a result of color information being confused as luminance information by the decoder circuits.

Chroma gain (chroma, color, saturation): In video, the gain of an amplifier as it pertains to the intensity of colors in the active picture.

Chroma key (color key): A video key effect in which one video signal is inserted in place of areas of a particular color in another video signal.

Chrominance: The color information of a color video signal.

Chrominance-to-luminance intermodulation (cross-talk, cross-modulation): An undesirable change in luminance amplitude caused by superimposition of some chrominance information on the luminance signal. Appears in a TV picture as unwarranted brightness variations caused by changes in color saturation levels.

CIE: Commission Internationale de l'Eclairagé. This is the International Committee for Light, established in 1965. It defines and recommends light units.

Cladding: The outer part of a fiber-optic cable, which is also a fiber but with a smaller material density than the center core. It enables a total reflection effect so that the light transmitted through the internal core stays inside.

Clamping (DC): The circuit or process that restores the DC component of a signal. A video clamp circuit, usually triggered by horizontal synchronizing pulses, reestablishes a fixed DC reference level for the video signal. A major benefit of a clamp is the removal of low-frequency interference, especially power line hum.

Clipping level: An electronic limit to avoid overdriving the video portion of the television signal.

C-mount: The first standard for CCTV lens screw mounting. It is defined with the thread of 1 inch (2.54 mm) in diameter and 32 threads per inch, and the back flange-to-CCD distance of 17.526 mm (0.69 inches). The C-mount description applies to both lenses and cameras. C-mount lenses can be put on both C-mount and CS-mount cameras, only in the latter case an adaptor is required.

CMYK: A color-encoding system used by printers in which colors are expressed by the subtractive primaries (i.e., cyan, magenta, and yellow) plus black (called K). The black layer is added to give increased contrast and range on printing presses.

Coaxial cable: Also called *coax*. The most common type of cable used for copper transmission of video signals. It has a coaxial cross-section, where the center core is the signal conductor, while the outer shield protects it from external electromagnetic interference.

CODEC: Coder/Decoder. An encoder plus a decoder is an electronic device that compresses and decompresses digital signals. CODECs usually perform analog-to-digital and digital-to-analog conversion.

Color bars: A pattern generated by a video test generator, consisting of eight equal-width color bars. Colors are white (75 percent); black (7.5 percent setup level); 75 percent saturated pure colors red, green, and blue; and 75 percent saturated hues of yellow, cyan, and magenta (mixtures of two colors in 1:1 ratio without third color).

Color carrier: The subfrequency in a color video signal (4.43 MHz for PAL) that is modulated with the color information. The color carrier frequency is chosen so that its spectrum interleaves with the luminance spectrum with minimum interference.

Color difference signal: A video color signal created by subtracting luminance and/or color information from one of the primary color signals (red, green, or blue). In the Betacam color difference format, for example, the luminance (Y) and color difference components (R–Y and B–Y) are derived as follows:

$$Y = 0.3 \quad \text{Red} + 0.59 \quad \text{Green} + 0.11 \quad \text{Blue}$$

$$R{-}Y = 0.7 \quad \text{Red} - 0.59 \quad \text{Green} - 0.11 \quad \text{Blue}$$

$$B{-}Y = 0.89 \quad \text{Blue} - 0.59 \quad \text{Green} - 0.3 \quad \text{Red}$$

The G-V color difference signal is not created, because it can be reconstructed from the other three signals. Other color difference conventions include SMPTE, EBU-N10, and MII. Color difference signals should not be referred to as *component video signals*. That term is reserved for the RGB color components. In informal use, the term *component video* is often used to mean color difference signals.

Color field: In the NTSC system, the color subcarrier is phase-locked to the line sync so that on each consecutive line, subcarrier phase is changed 180 degrees with respect to the sync pulses. In the PAL system, color subcarrier phase moves 90 degrees every frame. In NTSC this creates four different field types, whereas in PAL there are eight. In order to make clean edits, alignment of color field sequences from different sources is crucial.

Color frame: In color television, four (NTSC) or eight (PAL) properly sequenced color fields compose one color frame.

Color phase: The timing relationship in a video signal that is measured in degrees and keeps the hue of a color signal correct.

Color subcarrier: The 3.58 MHz signal that carries color information. This signal is superimposed on the luminance level. Amplitude of the color subcarrier represents saturation and phase angle represents hue.

Color temperature: Indicates the hue of the color. It is derived from photography where the spectrum of colors is based on a comparison of the hues produced when a black body (as in physics) is heated from red through yellow to blue, which is the hottest. Color temperature measurements are expressed in Kelvin.

Comb filter: An electrical filter circuit that passes a series of frequencies and rejects the frequencies in between, producing a frequency response similar to the teeth of a comb. Used on encoded video to select the chrominance signal and reject the luminance signal, thereby reducing cross-chrominance artifacts, or, conversely, to select the luminance signal and reject the chrominance signal, thereby reducing cross-luminance artifacts. Introduced in the S-VHS concept for a better luminance resolution.

Composite sync: A signal consisting of horizontal sync pulses, vertical sync pulses, and equalizing pulses only, with a no-signal reference level.

Composite video signal: A signal in which the luminance and chrominance information have been combined using one of the coding standards—NTSC, PAL, SECAM, and so on.

Concave lens: A lens that has negative focal length (i.e., the focus is virtual and it reduces the objects).

Contrast: A common term used in reference to the video picture dynamic range (i.e., the difference between the darkest and the brightest parts of an image).

Convex lens: A convex lens has a positive focal length (i.e., the focus is real). It is usually called a magnifying glass because it magnifies objects.

CPU: Central processing unit. A common term used in computers.

CRO: Cathode ray oscilloscope (see *oscilloscope*).

Cross-talk: A type of interference or undesired transmission of signals from one circuit into another circuit in the same system. Usually caused by unintentional capacitance (AC coupling).

CS-mount: A newer standard for lens mounting. It uses the same physical thread as the C-mount, but the back flange-to-CCD distance is reduced to 12.5 mm in order to have the lenses made smaller, more compact, and less expensive. CS-mount lenses can only be used on CS-mount cameras.

CS-to-C-mount adaptor: An adaptor used to convert a CS-mount camera to a C-mount to accommodate a C-mount lens. It looks like a ring 5 mm thick, with a male thread on one side and a female on the other, with a 1-inch diameter and 32 threads per inch. It usually comes packaged with the newer type (CS-mount) of cameras.

CVBS: Composite video bar signal. In broadcast television, this refers to the video signal, including the color information and syncs.

D/A (also DA): Opposite of A/D (i.e., digital-to-analog conversion).

Dark current: Leakage signal from a CCD sensor in the absence of incident light.

Dark noise: Noise caused by the random (quantum) nature of the dark current.

DAT: Digital audiotape. A system developed initially for recording and playback of digitized audio signals, maintaining signal quality equal to that of a CD. Recent developments in hardware and software might lead to a similar inexpensive system for video archiving, recording, and playback.

db: Decibel. A logarithmic ratio of two signals or values, this usually refers to power, but can also refer to voltage and current. When power is calculated, the logarithm is multiplied by 10, whereas for current and voltage it is multiplied by 20.

DBS: Direct broadcast satellite. Broadcasting from a satellite directly to a consumer user, usually using a small-aperture antenna.

DC: Direct current. Current that flows in only one direction, as opposed to AC.

DCT: Discrete cosine transform. Mathematical algorithm used to generate frequency representations of a block of video pixels. The DCT is an invertible, discrete orthogonal transformation between time and frequency domain. It can be either forward discrete cosine transform (FDCT) or inverse discrete cosine transform (IDCT).

Decoder: A device used to recover the component signals from a composite (encoded) source.

Degauss: To demagnetize.

Delay line: An artificial or real transmission line or equivalent device designed to delay a wave or signal for a specific length of time.

Demodulator: A device that strips the video and audio signals from the carrier frequency.

Depth of field: The area in front of and behind the object in focus that appears sharp on the screen. The depth of field increases with the decrease of the focal length (i.e., the shorter the focal length, the wider the depth of field). The depth of field is always wider behind objects in focus.

Dielectric: An insulating (nonconductive) material.

Differential gain: A change in subcarrier amplitude of a video signal caused by a change in luminance level of the signal. The resulting TV picture will show a change in color saturation caused by a simultaneous change in picture brightness.

Differential phase: A change in the subcarrier phase of a video signal caused by a change in the luminance level of the signal. The hue of colors in a scene change with the brightness of the scene.

Digital disk recorder: A system that allows recording of video images on a digital disk.

Digital signal: An electronic signal where every different value from the real-life excitation (sound, light) has a different value of binary combinations (words) that represent the analog signal.

DIN: Deutsche Industrie-Normen. Germany's standard.

Disk: A flat, circular plate, coated with a magnetic material, on which data may be stored by selective magnetization of portions of the surface. May be a flexible, floppy disk or a rigid, hard disk. It could also be a plastic compact disc (CD) or digital video disk (DVD).

Distortion: Nonproportional representation of an original.

DMD: Digital micro-mirror device. A new video projection technology that uses chips with a large number of miniature mirrors, whose projection angle can be controlled with digital precision.

DOS: Disk operating system. A software package that makes a computer work with its hardware devices, such as hard drive, floppy drive, screen, keyboard, and so on.

Dot pitch: The distance in millimeters between individual dots on a monitor screen. The smaller the dot pitch, the better because it allows for more dots to be displayed and better resolution. The dot pitch defines the resolution of a monitor. A high-resolution CCTV or computer monitor would have a dot pitch of less than 0.3 mm.

Drop-frame time code: SMPTE time code format that continuously counts 30 frames per second, but drops two frames from the count every minute except for every tenth minute (drops 108 frames every hour) to maintain synchronization of time code with clock time. This is necessary because the actual frame rate of NTSC video is 29.94 frames per second rather than an even 30 frames.

DSP: Digital signal processing. It usually refers to the electronic circuit section of a device capable of processing digital signals.

Dubbing: Transcribing from one recording medium to another.

Duplex: A communication system that carries information in both directions is called a *duplex system*. In CCTV, duplex is often used to describe the type of multiplexer that can perform two functions simultaneously: recording in multiplex mode and playback in multiplex mode. It can also refer to duplex communication between a matrix switcher and a PTZ site driver, for example.

D-VHS: A new standard proposed by JVC for recording digital signals on a VHS video recorder.

DV-mini: Mini digital video. A new format for audio and video recording on small camcorders, adopted by most camcorder manufacturers. Video and sound are recorded in a digital format on a small cassette (66 x 48 x 12 mm), superseding S-VHS and Hi 8 quality.

Dynamic range: The difference between the smallest amount and the largest amount that a system can represent.

EBU: European Broadcasting Union.

EIA: Electronics Industry Association, which has recommended the television standard used in the United States, Canada, and Japan, based on 525-line interlaced scanning. Formerly known as RMA or RETMA.

Encoder: A device that superimposes electronic signal information on other electronic signals.

Encryption: The rearrangement of the bitstream of a previously digitally encoded signal in a systematic fashion to make the information unrecognizable until restored on receipt of the necessary authorization key. This technique is used for securing information transmitted over a communication channel with the intent of excluding all but authorized receivers from interpreting the message. Can be used for voice, video, and other communications signals.

ENG camera: Electronic news gathering camera. Refers to CCD cameras in the broadcast industry.

EPROM: Erasable and programmable read-only memory. An electronic chip used in many different security products that stores software instructions for performing various operations.

Equalizer: Equipment designed to compensate for loss and delay frequency effects within a system. A component or circuit that allows for adjustment of a signal across a given band.

Ethernet: A local area network (LAN) used for connecting computers, printers, workstations, terminals, and so on. within the same building. Ethernet operates over twisted wire and coaxial cable at speeds up to 10 Mbps. Ethernet specifies a carrier sense multiple access with collision detection (CSMA/CD), which is a technique of sharing a common medium (wire, coaxial cable) among several devices.

External synchronization: A means of ensuring that all equipment is synchronized to the one source.

FCC: Federal Communications Commission (United States).

FFT: Fast Fourier transformation.

Fiber optics: A technology designed to transmit signals in the form of pulses of light. Fiber-optic cable is noted for its properties of electrical isolation and resistance to electrostatic and electromagnetic interference.

Field: Refers to one-half of the TV frame that is composed of either all odd or all even lines. In CCIR systems each field is composed of $625 \div 2 = 312.5$ lines; in EIA systems, $525 \div 2 = 262.5$ lines. There are 50 fields per second in CCIR/PAL, and 60 in the EIA/NTSC TV system.

Film recorder: A device for converting digital data into film output. Continuous-tone recorders produce color photographs as transparencies, prints, or negatives.

Fixed focal-length lens. A lens with a predetermined fixed focal length, a focusing control, and a choice of iris functions.

Flash memory: Nonvolatile, digital storage. Flash memory has slower access than SRAM or DRAM.

Flicker: An annoying picture distortion, mainly related to vertical syncs and video fields display. Some flicker normally exists as a result of interlacing; more apparent in 50-Hz systems (PAL). Flicker also shows when static images are displayed on the screen, such as computer-generated text transferred to video. Poor digital image treatment, found in low-quality system converters (going from PAL to NTSC and vice versa), creates an annoying flicker on the screen. There are several electronic methods to minimize flicker.

F-number: In lenses with adjustable irises, the maximum iris opening is expressed as a ratio (focal length of the lens)/(maximum diameter of aperture). This maximum iris is engraved on the front ring of the lens.

Focal length: The distance between the optical center of a lens and the principal convergent focus point.

Focusing control: A means of adjusting the lens to allow objects at various distances from the camera to be sharply defined.

Foot-candela: An illumination light unit used mostly in American CCTV terminology. It equals 10 times (more precisely, 9.29 times) the illumination value in Luxes.

Fourier transformation: Mathematical transformation of time domain functions into frequency domain.

Frame: (See also *field*.) Refers to a composition of lines that make one TV frame. In CCIR/ PAL TV system one frame is composed of 625 lines, whereas in EIA/

NTSC TV systems one frame is composed of 525 lines. There are 25 frames per second in the CCIR/PAL and 30 in the EIA/NTSC TV systems.

Frame-interline transfer (FIT): Refers to one of the three principles of charge transfer in CCD chips. The other two are interline and frame transfer.

Frame store: An electronic device that digitizes a TV frame (or TV field) of a video signal and stores it in memory. Multiplexers, fast scan transmitters, quad compressors, and even some of the latest color cameras have built-in frame stores.

Frame switcher: Another name for a simple multiplexer, which can record multiple cameras on a single VCR (and play back any camera in full screen) but does not have a mosaic image display.

Frame synchronizer: A digital buffer that, by storage and comparison of sync information to a reference and timed release of video signals, can continuously adjust the signal for any timing errors.

Frame transfer (FT): Refers to one of the three principles of charge transfer in CCD chips. The other two are interline and frame-interline transfer.

Frequency: The number of complete cycles of a periodic waveform that occur in a given length of time. Usually specified in cycles per second (Hz).

Frequency modulation (FM): Modulation of a sine wave or carrier by varying its frequency in accordance with amplitude variations of the modulating signal.

Front porch: The blanking signal portion that lies between the end of the active picture information and the leading edge of horizontal sync.

Gain: Any increase or decrease in strength of an electrical signal. Gain is measured in terms of decibels or number of times of magnification.

Gamma: A correction of the linear response of a camera in order to compensate for the monitor phosphor screen nonlinear response. It is measured with the exponential value of the curve describing the nonlinearity. A typical monochrome monitor's gamma is 2.2, and a camera needs to be set to the inverse value of 2.2 (which is 0.45) for the overall system to respond linearly (i.e., unity).

Gamut: The range of voltages allowed for a video signal or a component of a video signal. Signal voltages outside of the range (i.e., exceeding the gamut) may lead to clipping, cross talk, or other distortions.

GB: Gigabyte. Unit of computer memory consisting of about 1,000 million bytes (1,000 megabytes). Actual value is 1,073,741,824 bytes.

Gen-lock: A way of locking the video signal of a camera to an external generator of synchronization pulses.

GHz: Gigahertz. One billion cycles per second.

GND: Ground (electrical).

Gray scale: A series of tones that range from true black to true white, usually expressed in ten steps.

Ground loop: An unwanted interference in the copper electrical signal transmissions with shielded cable, which is a result of ground currents when the system has more than one ground—for example, in CCTV, when we have a different grounding resistance at the camera than at the switcher or monitor end. The induced electrical noise generated by the surrounding electrical equipment (including mains) does not discharge equally through the two groundings (because they are different), and the induced noise shows up on the monitors as interference.

GUI: Graphical user interface.

HAD: Hole-accumulated diode. A type of CCD sensor with a layer designed to accumulate holes (in the electronic sense), thus reducing noise level.

HDD: Hard disk drive. A magnetic medium for storing digital information on most computers and electronic equipment that process digital data.

HDDTV: High-definition digital television. The upcoming standard of broadcast television with extremely high resolution and an aspect ratio of 16:9. It is an advancement from the analog high definition, already used experimentally in Japan and Europe. The picture resolution is nearly 2,000 × 1,000 pixels, and it uses the MPEG-2 standard.

HDTV: High-definition television. It usually refers to the analog version of HDDTV. The SMPTE in the United States and ETA in Japan have proposed an HDTV product standard: 1,125 lines at 60 Hz field rate, 2:1 interlace; 16:9 aspect ratio; 30 MHz RGB and luminance bandwidth.

Headend: The electronic equipment located at the start of a cable television system, usually including antennas, earth stations, preamplifiers, frequency converters, demodulators, modulators, and related equipment.

Helical scan: A method of recording video information on a tape, most commonly used in home and professional VCRs.

Herringbone: Patterning caused by driving a color-modulated composite video signal (PAL or NTSC) into a monochrome monitor.

Hertz: A unit that measures the number of certain oscillations per second.

Horizontal drive (also horizontal sync): This signal is derived by dividing the subcarrier by 227.5 and then doing some pulse shaping. The signal is used by monitors and cameras to determine the start of each horizontal line.

Horizontal resolution: Chrominance and luminance resolution (detail) expressed horizontally across a picture tube. This is usually expressed as a number of black-to-white transitions or lines that can be differentiated. Limited by the bandwidth of the video signal or equipment.

Horizontal retrace: At the end of each horizontal line of video, a brief period when the scanning beam returns to the other side of the screen to start a new line.

Horizontal sync pulse: The synchronizing pulse at the end of each video line that determines the start of horizontal retrace.

Housings, environmental: Usually refers to containers for cameras and lenses and associated accessories, such as heaters, washers, and wipers, to meet specific environmental conditions.

HS: Horizontal sync.

Hue (tint, phase, chroma phase): One of the characteristics that distinguishes one color from another. Hue defines color on the basis of its position in the spectrum (i.e., whether red, blue, green, or yellow). Hue is one of the three characteristics of television color (see also *saturation* and *luminance*). In NTSC and PAL video signals, the hue information at any particular point in the picture is conveyed by the corresponding instantaneous phase of the active video subcarrier.

Hum: A term used to describe an unwanted induction of mains frequency.

Hum bug: Another name for a ground loop corrector.

Hyper-HAD. An improved version of the CCD HAD technology, utilizing on-chip micro lens technology to provide increased sensitivity without increasing the pixel size.

IDE: Interface device electronics. Software and hardware communication standard for interconnecting peripheral devices to a computer.

IEC: International Electrotechnical Commission (also CEI).

Imaging device: A vacuum tube or solid-state device in which the vacuum tube light-sensitive face plate or solid-state light-sensitive array provides an electronic signal from which an image can be created.

Impedance: A property of all metallic and electrical conductors that describes the total opposition to current flow in an electrical circuit. Resistance, inductance, capacitance, and conductance have various influences on the impedance, depending on frequency, dielectric material around conductors, physical relationship between conductors and external factors. Impedance is often referred to with the letter Z. It is measured in ohms, whose symbol is the Greek letter omega, Ω.

Input: Same as I/P.

Inserter (also alphanumeric video generator): A device for providing additional information, normally superimposed on the picture being displayed; this can range from one or two characters to full-screen alphanumeric text. Usually, such generators use the incoming video signal sync pulses as a reference point for the text insertion position, which means, if the video signal is of poor quality, the text stability will also be of poor quality.

Interference: Disturbances of an electrical or electromagnetic nature that introduce undesirable responses in other electronic equipment.

Interlaced scanning: A technique of combining two television fields in order to produce a full frame. The two fields are composed of only odd and only even lines, which are displayed one after the other but with the physical position of all the lines interleaving each other—hence, interlace. This type of television picture creation was proposed in the early days of television to have a minimum amount of information yet achieve flickerless motion.

Interline transfer: This refers to one of the three principles of charge transferring in CCD chips. The other two are frame transfer and frame-interline transfer.

I/O: Input/Output.

IP: Index of protection. A numbering system that describes the quality of protection of an enclosure from outside influences, such as moisture, dust, and impact.

I/P: Input. A signal applied to a piece of electric apparatus or the terminals on the apparatus to which a signal or power is applied.

I^2R: Formula for power in watts (W), where I is current in amperes (A), and R is resistance in ohms (W).

IRE: Institute of Radio Engineers. Units of measurement dividing the area from the bottom of sync to peak white level into 140 equal units. A total of 140 IRE equals 1 Vpp. The range of active video is 100 IRE.

Iris: A means of controlling the size of a lens aperture and therefore the amount of light passing through the lens.

IR light: Infrared light, invisible to the human eye. It usually refers to wavelengths longer than 700 nm. Monochrome (b/w) cameras have extremely high sensitivity in the infrared region of the light spectrum.

ISDN: Integrated services digital network. The newer generation telephone network, which uses 64 Kbps speed of transmission (being a digital network, the signal bandwidth is not expressed in kHz but rather with a transmission speed). This is much faster than a normal PSTN telephone line. To use the ISDN network, you have to talk to your communications provider, but in general a special set of interface units (such as modems) are required.

ISO: International Organization for Standardization.

ITU. International Telecommunications Union.

JPEG: Joint Photographic Experts Group. A group that has recommended a compression algorithm for still digital images that can compress with ratios of over 10:1. Also the name of the format itself.

kb/s: Kilobits per second. Thousand bits per second. Also written as kbps.

Kelvin: One of the basic physical units of measurement for temperature. The scale is the same as the Celsius, but the 0°K starts from –273°C. Also, the unit of measurement of the temperature of light is expressed in Kelvins or K. In color recording, light temperature affects the color values of the lights and the scene that they illuminate.

K factor: A specification rating method that gives a higher factor to video disturbances that cause the most observable picture degradation.

kHz: Kilohertz. 1,000 Hertz.

Kilobaud: A unit of measurement of data transmission speed equaling 1,000 baud.

Kilobyte: 1,024 bytes.

Lambertian source or surface: A surface is called a Lambert radiator or reflector (depending on whether the surface is a primary or a secondary source of light) if it is a perfectly diffusing surface.

LAN: Local area network. A short-distance data communications network (typically within a building or campus) used to link together computers and peripheral devices (such as printers, CD-ROMs, and modems) under some form of standard control.

Laser: Light amplification by stimulated emission of radiation. A laser produces a very strong and coherent light of a single frequency.

LED: Light-emitting diode. A semiconductor that produces light when a certain low voltage is applied to it in one direction.

Lens: An optical device for focusing a desired scene onto the imaging device in a CCTV camera.

Level: When relating to a video signal, it refers to the video level in volts. In CCTV optics, it refers to the auto-iris level setting of the electronics that processes the video signal in order to open or close the iris.

Line-locked: In CCTV, this usually refers to multiple cameras being powered by a common alternative current (AC) source (either 24 V AC, 110 V AC, or 240 V AC) and consequently has field frequencies locked to the same AC source frequency (50 Hz in CCIR systems and 60 Hz in EIA systems).

Liquid crystal display (LCD): A screen for displaying text/graphics based on a technology called liquid crystal, where minute currents change the reflectiveness or transparency of the screen. The advantages of LCD screens are small power consumption (can be easily battery driven) and low price of mass-produced units. The disadvantages are narrow viewing angle, slow response (a bit too slow to be used for video), invisibility in the dark unless the display is back-lighted, and difficulties displaying true colors with color LCD displays.

Lumen (lm): A light intensity produced by the luminosity of one candela in one radian of a solid angle.

Luminance: Refers to the video signal information about the scene brightness. The measurable luminous intensity of a video signal. Differentiated from brightness in that the latter is nonmeasurable and sensory. The color video picture information contains two components: luminance (brightness and contrast) and chrominance (hue and saturation). The photometric quantity of light radiation.

LUT: Look-up table. A cross-reference table in the computer memory that transforms raw information from the scanner or computer and corrects values to compensate for weakness in equipment or for differences in emulsion types.

Lux (lx): Light unit for measuring illumination. It is defined as the illumination of a surface when luminous flux of one lumen falls on an area of 1 m². It is also known as lumen per square meter or meter-candelas.

MAC: Multiplexed analog components. A system in which the components are time multiplexed into one channel using time domain techniques (i.e., the components are kept separate by being sent at different times through the same channel). There are many different MAC formats and standards.

Manual iris: A manual method of varying the size of a lens's aperture.

Matrix: A logical network configured in a rectangular array of intersections of input/output channels.

Matrix switcher: A device for switching more than one camera, VCR, video printer, and similar to more than one monitor, VCR, video printer, and similar. Much more complex and more powerful than video switchers.

MATV: Master antenna television.

MB: Megabyte. Unit of measurement for computer memory consisting of approximately one million bytes. Actual value is 1,048,576 bytes. Kilobyte × kilobyte = megabyte.

MB/s: Megabytes per second. Million bytes per second or 8 million hits per second. Also written as MBps.

MHz: Megahertz. One million hertz.

Microwave: One definition refers to the portion of the electromagnetic spectrum that ranges between 300 MHz and 3,000 GHz. The other definition is when referring to the transmission medium where microwave links are used. Frequencies in microwave transmission are usually between 1 GHz and 12 GHz.

MOD: Minimum object distance. Feature of a fixed or zoom lens that indicates the closest distance an object can be from the lens's image plane, expressed in meters. Zoom lenses have MOD of approximately 1 m, while fixed lenses usually have much less, depending on the focal length.

Modem: This popular term is made up of two words: modulate and demodulate. The function of a modem is to connect a device (usually a computer) via a telephone line to another device with a modem.

Modulation: The process by which some characteristic (i.e., amplitude, phase) of one RF wave is varied in accordance with another wave (message signal).

Moiré pattern: An unwanted effect that appears in the video picture when a high-frequency pattern is looked at with a CCD camera that has a pixel pattern close (but lower) than the object pattern.

Monochrome: Black-and-white video. A video signal that represents the brightness values (luminance) in the picture but not the color values (chrominance).

MPEG: Motion Picture Experts Group. An ISO group of experts that has recommended manipulation of digital motion images. Today there are a few MPEG recommendations, of which the most well known are MPEG-1 and MPEG-2. The latter is widely accepted for high-definition digital television, as well as multimedia presentation.

MPEG-1: Standard for compressing progressive scanned images with audio. Bit rate is from 1.5 Mbps up to 3.5 Mbps.

MPEG-2: The standard for compression of progressive scanned and interlaced video signals with high-quality audio over a large range of compression rates, with a range of bit rates from 1.5 to 100 Mbps. Accepted as an HDTV and DVD standard of video/audio encoding.

Noise: An unwanted signal produced by all electrical circuits working above the absolute zero. Noise cannot be eliminated but only minimized.

Non-drop-frame time code: SMPTE time code format that continuously counts a full 30 frames per second. Because NTSC video does not operate at exactly 30 frames per second, non-drop-frame time code will count 108 more frames in one hour than actually occur in the NTSC video in one hour. The result is incorrect synchronization of time code with clock time. Drop-frame time code solves this problem by skipping or dropping two frame numbers per minute, except at the tens of the minute count.

Noninterlaced: The process of scanning, whereby every line in the picture is scanned during the vertical sweep.

NTSC: National Television System Committee. American committee that set the standards for color television as used today in the United States, Canada, Japan, and parts of South America. NTSC television uses a 3.57945 MHz subcarrier, whose phase varies with the instantaneous hue of the televised color and whose amplitude varies with the instantaneous saturation of the color. NTSC employs 525 lines per frame and 59.94 fields per second.

Numerical aperture: A number that defines the light-gathering ability of a specific fiber. The numerical aperture is equal to the sine of the maximum acceptance angle.

Objective: The first optical element at the front of a lens.

Ocular: The last optical element at the back of a lens (the one closer to the CCD chip).

Ohm: The unit of resistance. The electrical resistance between two points of a conductor where a constant difference of potential of 1 V applied between these points produces in the conductor a current of 1 A, the conductor not being the source of any electromotive force.

O/P: Output.

Oscilloscope (also CRO, from cathode ray oscilloscope): An electronic device that can measure the signal changes versus time. A must for any CCTV technician.

Output impedance: The impedance a device presents to its load. The impedance measured at the output terminals of a transducer with the load disconnected and all impressed driving forces taken as zero.

Overscan: A video monitor condition in which the raster extends slightly beyond the physical edges of the CRT screen, cutting off the outer edges of the picture.

PAL: Phase alternating line. Describes the color phase change in a PAL color signal. PAL is a European color TV system featuring 625 lines per frame, 50 fields per second, and a 4.43361875-MHz subcarrier. Used mainly in Europe, China, Malaysia, Australia, New Zealand, the Middle East, and parts of Africa. PAL-M is a Brazilian color TV system with phase alternation by line, but using 525 lines per frame, 60 fields per second, and a 3.57561149-MHz subcarrier.

Pan and tilt head (P/T head): A motorized unit permitting vertical and horizontal positioning of a camera and lens combination. Usually 24 V AC motors are used in such P/T heads, but also 110 V AC (i.e., 240 V AC) units can be ordered.

Pan unit: A motorized unit permitting horizontal positioning of a camera.

Peak-to-peak (pp): The amplitude (voltage) difference between the most positive and the most negative excursions (peaks) of an electrical signal.

Pedestal: In the video waveform, the signal level corresponding to black. Also called *setup.*

Phase-locked loop (PLL): A circuit containing an oscillator whose output phase or frequency locks onto and tracks the phase or frequency of a reference input signal. To produce the locked condition, the circuit detects any phase difference between the two signals and generates a correction voltage that is applied to the oscillator to adjust its phase or frequency.

Phot: A photometric light unit for strong illumination levels. One phot is equal to 10,000 Luxes.

Photodiode: A type of semiconductor device in which a PN junction diode acts as a photosensor.

Photo-effect: Also known as photoelectric-effect. This refers to a phenomenon of ejection of electrons from a metal whose surface is exposed to light.

Photo multiplier: A highly light-sensitive device. Advantages are its fast response, good signal-to-noise ratio, and wide dynamic range. Disadvantages are fragility (vacuum tube), high voltage, and sensitivity to interference.

Photon: A representative of the quantum nature of light. It is considered to be the smallest unit of light.

Photopic vision: The range of light intensities, from 10^5 Lux down to nearly 10^{-2} Lux, detectable by the human eye.

Pinhole lens: A fixed focal-length lens, for viewing through a small aperture, used in discrete surveillance situations. The lens normally has no focusing control but offers a choice of iris functions.

Pixel: Derived from picture element. Usually refers to the CCD chip unit picture cell. It consists of a photosensor plus its associated control gates. The smallest visual unit that is handled in a raster file, generally a single cell in a grid of numbers describing an image.

Plumbicon: Thermionic vacuum tube developed by Philips, using a lead oxide photoconductive layer. It represented the ultimate imaging device until the introduction of CCD chips.

Polarizing filter: An optical filter that transmits light in only one direction (perpendicular to the light path) out of 360 degrees possible. The effect is such that it can eliminate some unwanted bright areas or reflections, such as when looking through a glass window. In photography, polarizing filters are often used to darken a blue sky.

POTS: Plain old telephone service. The telephone service in common use throughout the world today. Also known as *PSTN*.

P-picture: Prediction-coded picture. An MPEG term to describe a picture that is coded using motion-compensated prediction from the past reference picture.

Preset positioning: A function of a pan and tilt unit, including the zoom lens, where a number of certain viewing positions can be stored in the system's memory (usually this is in the PTZ site driver) and recalled when required, either on an alarm trigger, programmed, or by manual recall.

Primary colors: A small group of colors that, when combined, can produce a broad spectrum of other colors. In television, red, green, and blue are the primary colors from which all other colors in the picture are derived.

Principal point: One of the two points that each lens has along the optical axis. The principal point closer to the imaging device (CCD chip in our case) is used as a reference point when measuring the focal length of a lens.

PROM: Programmable read-only memory. A ROM that can be programmed by the equipment manufacturer (rather than the PROM manufacturer).

Protocol: A specific set of rules, procedures, or conventions relating to format and timing of data transmission between two devices. A standard procedure that two data devices must accept and use to be able to understand each other. The protocols for data communications cover such things as framing, error handling, transparency, and line control.

PSTN: Public switched telephone network. Usually refers to the plain old telephone service, also known as *POTS*.

PTZ camera: Pan/tilt and zoom camera.

PTZ site driver (or receiver or decoder): An electronic device, usually a part of a video matrix switcher, that receives digital encoded control signals in order to operate pan, tilt, zoom, and focus functions.

Pulse: A current or voltage that changes abruptly from one value to another and back to the original value in a finite length of time. Used to describe one particular variation in a series of wave motions.

QAM: Quadrature amplitude modulation. Method for modulating two carriers. The carriers can be analog or digital.

Quad compressor (also split-screen unit): Equipment that simultaneously displays parts or more than one image on a single monitor. It usually refers to a four-quadrant display.

Radio frequency (RF): A term used to describe incoming radio signals to a receiver or outgoing signals from a radio transmitter (above 150 Hz). Even though they are not properly radio signals, TV signals are included in this category.

RAID: Redundant arrays of independent disks. This technology connects several hard drives into one mass storage device, which can be used, among other things, for digital recording of video images.

RAM: Random access memory. The capacity (measured in kilobytes) of electronic chips, usually known as memory, that hold digital information while there is power applied. This is the computer's work area.

Random interlace: In a camera, a free-running horizontal sync as opposed to a 2:1 interlace type that has the sync locked and therefore has both fields in a frame interlocked together accurately.

Registration: An adjustment associated with color sets and projection TVs to ensure that the electron beams of the three primary colors of the phosphor screen are hitting the proper color dots/stripes.

Remote control: A transmitting and receiving of signals for controlling remote devices such as pan/tilt units, lens functions, wash and wipe control, and similar.

Resolution: A measure of the ability of a camera or television system to reproduce detail. The number of picture elements that can be reproduced with good definition.

RETMA: Former name of the EIA association. Some older video test charts carry the name RETMA Chart.

Retrace: The return of the electron beam in a CRT to the starting point after scanning. During retrace, the beam is typically turned off. All of the sync information is placed in this invisible portion of the video signal. May refer to retrace after each horizontal line or after each vertical scan (field).

RF signal: Radio frequency signal that belongs to the region up to 300 GHz.

RG-11: A video coaxial cable with 75-W impedance and of much thicker diameter than the popular RG-59 (of approximately 12 mm). With RG-11, much longer distances can be achieved (at least twice the RG-59), but it is more expensive and harder to work with.

RG-58: A coaxial cable designed with 50-W impedance; therefore, not suitable for CCTV. Very similar to RG-59, only slightly thinner.

RG-59: A type of coaxial cable that is most common in use in small- to medium-size CCTV systems. It is designed with an impedance of 75 W. It has an outer diameter of approximately 6 mm, and it is a good compromise between maximum distances achievable (up to 300 m for monochrome signals and 250 m for color) and good transmission.

Rise time: The time taken for a signal to make a transition from one state to another; usually measured between the 10 and 90 percent completion points of the transition. Shorter or faster rise times require more bandwidth in a transmission channel.

RMS: Root mean square. A measure of effective (as opposed to peak) voltage of an AC waveform. For a sine wave it is 0.707 times the peak voltage. For any periodic waveform, it is the square root of the average of the squares of the values through one cycle.

ROM: Read-only memory. An electronic chip containing digital information that does not disappear when power is turned off.

Routing switcher: An electronic device that routes a user-supplied signal (e.g., audio, video) from any input to any user-selected output. This is a broadcast term for matrix switchers, as we know them in CCTV.

RS-125: An SMPTE parallel component digital video standard.

RS-170: A document prepared by the Electronics Industries Association describing recommended practices for NTSC color television signals in the United States.

RS-232: A format of digital communication where only two wires are required. It is also known as a *serial data communication*. The RS-232 standard defines a scheme for asynchronous communications, but it does not define how the data should be represented by the bits (i.e., it does not define the overall message format and protocol). It is often used in CCTV communications between keyboards and matrix switchers or between matrix switchers and PTZ site drivers. The advantage of RS-232 over others is its simplicity and use of only two wires.

RS-422: This is an advanced format of digital communication when compared with RS-232. The basic difference is in the need for four wires instead of two, because the communication is not single-ended as with RS-232 but differential. In simple terms, the signal transmitted is read at the receiving end as the difference between the two wires without common ground. So if there is noise induced along the line, it will be canceled out. The RS-422 can drive lines of more than one kilometer in length and distribute data to up to ten receivers.

RS-485: This is an advanced format of digital communications compared with RS-422. The major improvement is in the number of receivers that can be driven with this format, up to 32.

Saturation (in color): The intensity of the colors in the active picture. The degree by which the eye perceives a color as departing from a gray or white scale of the same brightness. A 100-percent saturated color does not contain any white; adding white reduces saturation. In NTSC and PAL video signals, the color saturation at any particular instant in the picture is conveyed by the corresponding instantaneous amplitude of the active video subcarrier.

Scanner: When referring to a CCTV device, it is the pan-only head. When referring to an imaging device, it is the device with a CCD chip that scans documents and images.

Scanning: The rapid movement of the electron beam in the CRT of a monitor or television receiver. It is formatted line for line across the photosensitive surface to produce or reproduce the video picture. When referring to a PTZ camera, it is the panning or horizontal camera motion.

Scene illumination: The average light level incident on a protected area. Normally measured for the visible spectrum with a light meter having a spectral response corresponding closely to that of the human eye and is quoted in Lux.

Scotopic vision: Illumination levels below 10^{-2} Lux, thus invisible to the human eye.

SCSI: Small computer systems interface. A computer standard that defines the software and hardware methods of connecting more external devices to a computer bus.

SECAM: Sequential couleur avec memoire (sequential color with memory). A color television system with 625 lines per frame (used to be 819) and 50 fields per second developed by France and the former U.S.S.R. Color difference information is transmitted sequentially on alternate lines as an FM signal.

Serial data: Time-sequential transmission of data along a single wire. In CCTV, the most common method of communicating between keyboards and the matrix switcher and also controlling PTZ cameras.

Serial interface: A digital communications interface in which data are transmitted and received sequentially along a single wire or pair of wires. Common serial interface standards are RS-232 and RS-422.

Serial port: A computer input/output (I/O) port through which the computer communicates with the external world. The standard serial port is RS-232 based and allows bidirectional communication on a relatively simple wire connection as data flow serially.

Sidebands: The frequency bands on both sides of a carrier within which the energy produced by the process of modulation is carried.

Signal-to-noise ratio (S/N): An S/N ratio can be given for the luminance signal, chrominance signal, and audio signal. The S/N ratio is the ratio of noise to actual total signal, and it shows how much higher the signal level is than the level of noise. It is expressed in decibels (db), and the bigger the value is, the crisper and clearer the picture and sound will be during playback. An S/N ratio is calculated with the logarithm of the normal signal and the noise RMS value.

Silicon: The material of which modern semiconductor devices are made.

Simplex: In general, it refers to a communications system that can transmit information in one direction only. In CCTV, simplex is used to describe a method of multiplexer operation where only one function can be performed at a time (e.g., either recording or playback individually).

Single-mode fiber: An optical glass fiber that consists of a core of very small diameter. A typical single-mode fiber used in CCTV has a 9-mm core and a 125-mm outer diameter. Single-mode fiber has less attenuation and therefore transmits signals at longer distances (up to 70 km). Such fibers are normally used only with laser sources because of their small acceptance cone.

Skin effect: The tendency of alternating current to travel only on the surface of a conductor as its frequency increases.

Slow scan: The transmission of a series of frozen images by means of analog or digital signals over limited bandwidth media, usually telephone.

Smear: An unwanted side effect of vertical charge transfer in a CCD chip. It shows vertical bright stripes in places of the image where there are bright areas. In better cameras, smear is minimized to almost undetectable levels.

SMPTE: Society of Motion Picture and Television Engineers.

SMPTE time code: In video editing, time code that conforms to SMPTE standards. It consists of an eight-digit number specifying hours: minutes: seconds: frames. Each number identifies one frame on a videotape. SMPTE time code may be of either the drop-frame or non-drop-frame type.

Snow: Random noise on the display screen, often resulting from dirty heads or weak broadcast video reception.

S/N ratio: See *signal-to-noise ratio.*

Spectrum: In electromagnetics, spectrum refers to the description of a signal's amplitude versus its frequency components. In optics, spectrum refers to the light frequencies composing the white light, which can be seen as rainbow colors.

Spectrum analyzer: An electronic device that can show the spectrum of an electric signal.

SPG: Sync pulse generator. A source of synchronization pulses.

Split-screen unit (quad compressor): Equipment that simultaneously displays different parts of an image or more than one image on a single monitor. It usually refers to four-quadrant display.

Staircase (in television): Same as color bars. A pattern generated by the TV generator, consisting of equal-width luminance steps of 0, +20, +40, +60, +80, and +100 IRE units and a constant-amplitude chroma signal at color burst phase. Chroma amplitude is selectable at 20 IRE units (low stairs) or 40 IRE units (high stairs). The staircase pattern is useful for checking linearity of luminance and chroma gain, differential gain, and differential phase.

Start bit: A bit preceding the group of bits representing a character used to signal the arrival of the character in asynchronous transmission.

Subcarrier (SC): These are the basic signals in all NTSC and PAL sync signals: 3.58 MHz for NTSC, 4.43 MHz for PAL. It is a continuous sine wave, usually generated and distributed at 2 V in amplitude and having a frequency of 3.579545 MHz (NTSC) and 4.43361875 MHz (PAL). Subcarrier is usually divided down from a primary crystal running at 14.318180 MHz, for example, in NTSC, and that divided by 4 is 3.579545. The formula is similar with PAL. All other synchronizing signals are directly divided down from subcarrier.

S-VHS: Super VHS format in video recording. A newer standard proposed by JVC, preserving the downward compatibility with the VHS format. It offers much better horizontal resolution of up to 400 TV lines, mainly because of the color separation techniques, high-quality video heads, and better tapes. S-VHS is usually associated with Y/C separated signals.

Sync: Short for synchronization pulse.

Sync generator (sync pulse generator, SPG): Device that generates synchronizing pulses needed by video source equipment to provide proper equipment video signal timing. Pulses typically produced by a sync generator could be subcarrier, burst flag, sync, blanking, H and V drives, and color black. Most commonly used in CCTV are H and V drives.

T1: A digital transmission link with a capacity of 1.544 Mbps. T1 uses two pairs of normal twisted wires. T1 lines are used for connecting networks across remote distances. Bridges and routers are used to connect LANs over T1 networks.

T1 channels: In North America, a digital transmission channel carrying data at a rate of 1.544 million bits per second. In Europe, a digital transmission channel carrying data at a rate of 2.048 million bits per second. AT&T term for a digital carrier facility used to transmit a DS-1 formatted digital signal at 1.544 Mbps.

T3 channels: In North America, a digital channel that communicates at 45.304 Mbps; commonly referred to by its service designation of DS-3.

TBC: Time base correction. Synchronization of various signals inside a device such as a multiplexer or a time base corrector.

TDG: Time and date generator.

TDM: Time division multiplex. A time-sharing of a transmission channel by assigning each user a dedicated segment of each transmission cycle.

Tearing: A lateral displacement of the video lines caused by sync instability. It appears as though parts of the images have been torn away.

Teleconferencing: Electronically linked meeting conducted among groups in separate geographical locations.

Telemetry: Remote controlling system of, usually, digital encoded data, intended to control pan, tilt, zoom, focus, preset positions, wash, wipe, and similar functions. Being digital, it is usually sent via twisted-pair cable or coaxial cable together with the video signal.

Termination: This usually refers to the physical act of terminating a cable with a special connector, which, for coaxial cable, is usually BNC. For fiber-optic cable, this is the ST connector. It can also refer to the impedance matching when electrical transmission is in use. This is especially important for high-frequency signals, such as the video signal, where the characteristic impedance is accepted to be 75 W.

TFT: Thin-film transistor. This technology is used mainly for manufacturing flat computer and video screens that are superior to the classic LCD screens. Color quality, fast response time, and resolution are excellent for video.

Time-lapse VCR (TL VCR): A video recorder, most often in VHS format, that can prolong the video recording on a single tape up to 960 hours (this refers to a 180-minute tape). This type of VCR is often used in CCTV systems. The principle of operation is simple: Instead of having the videotape travel at a constant speed of 2.275 cm/s (which is the case with the domestic models of VHS VCRs), it moves with discrete steps that can be controlled. Time-lapse VCRs have several other special functions that are useful in CCTV, such as external alarm trigger, time and date superimposed on the video signal, alarm search, and so on.

Time-lapse video recording: The intermittent recording of video signals at intervals to extend the recording time of the recording medium. It is usually measured in reference to a 3-hour (180-minute) tape.

Time multiplexing: The technique of recording several cameras onto one time-lapse VCR by sequentially sending camera pictures with a timed interval delay to match the time-lapse mode selected on the recorder.

T-pulse to bar: A term relating to frequency response of video equipment. A video signal containing equal amplitude T-pulse and bar portions is passed through the equipment, and the relative amplitudes of the T-pulse and bar are measured at the output. A loss of response is indicated when one portion of the signal is lower in amplitude than the other.

Tracking: The angle and speed at which the tape passes the video heads.

Transcoder: A device that converts one form of encoded video to another (e.g., to convert NTSC video to PAL). Sometimes mistakenly used to mean translator.

Transducer: A device that converts one form of energy into another—for example, in fiber optics, a device that converts light signals into electrical signals.

Translator: A device used to convert one component set to another (e.g., to convert Y, R-Y, B-Y signals to RGB signals).

Transponder: The electronics of a satellite that receives an uplinked signal from earth, amplifies it, converts it to a different frequency, and returns it to earth.

TTL: Transistor-transistor logic. A term used in digital electronics mainly to describe the ability of a device or circuit to be connected directly to the input or output of digital equipment. Such compatibility eliminates the need for interfacing circuitry. TTL signals are usually limited to two states, low and high, and are thus much more limited than analog signals. Also stands for through-the-lens viewing or color measuring.

Twisted-pair: A cable composed of two small, insulated conductors twisted together. Because both wires have nearly equal exposure to any interference, the differential noise is slight.

UHF signal: Ultra high frequency signal. In television it is defined to belong in the radio spectrum between 470 MHz and 850 MHz.

Unbalanced signal: In CCTV, this refers to a type of video signal transmission through a coaxial cable. It is called unbalanced because the signal travels through the center core only, while the cable shield is used for equating the two voltage potentials between the coaxial cable ends.

Underscan: Decreases raster size H and V so that all four edges of the picture are visible on the monitor.

UPS: Uninterruptible power supply. These are power supplies used in most high-security systems, whose purpose is to back up the system for at least 10 minutes without main power. The duration of this backup depends on the size of the UPS, usually expressed in VA, and the current consumption of the system itself.

UTP: Unshielded twisted pair. A cable medium with one or more pairs of twisted, insulated copper conductors bound in a single sheath. Now the most common method of bringing telephone and data to the desktop.

Variable bit rate: Operation where the bit rate varies with time during the decoding of a compressed bit stream.

VDA: See *video distribution amplifier.*

Vectorscope: An instrument similar to an oscilloscope that is used to check and/or align amplitude and phase of the three color signals (RGB).

Velocity of propagation: Speed of signal transmission. In free space, electromagnetic waves travel at the speed of light. In coaxial cables, this speed is reduced by the dielectric material. Commonly expressed as percentage of the speed in free space.

Vertical interval: The portion of the video signal that occurs between the end of one field and the beginning of the next. During this time, the electron beams in the monitors are turned off (invisible) so that they can return from the bottom of the screen to the top to begin another scan.

Vertical interval switcher: A sequential or matrix switcher that switches from one camera to another exactly in the vertical interval, thus producing roll-free switching. This is possible only if the various camera sources are synchronized.

Vertical resolution: Chrominance and luminance detail expressed vertically in the picture tube. Limited by the number of scan lines.

Vertical retrace: The return of the electron beam to the top of a television picture tube screen or a camera pickup device target at the completion of the field scan.

Vertical shift register: The mechanism in CCD technology whereby charge is read out from the photosensors of an interline transfer or frame-interline transfer sensor.

Vertical sync pulse: A portion of the vertical blanking interval that is made up of blanking level. Synchronizes vertical scan of television receiver to composite video signal. Starts each frame at same vertical position.

Vestigial sideband transmission: A system of transmission wherein the sideband on one side of the carrier is transmitted only in part.

VGA: Video graphics array.

VHF: Very high frequency. A signal encompassing frequencies between 30 and 300 MHz. In television, VHF Band I uses frequencies between 45 MHz and 67 MHz; Band II is reserved for FM radio from 88 MHz to 108 MHz; and Band III is between 180 MHz and 215 MHz.

VHS: Video home system. As proposed by JVC, a video recording format used most often in homes but also in CCTV. Its limitations include the speed of recording, the magnetic tapes used, and the color separation technique. Most of the CCTV equipment today supersedes VHS resolution.

Video bandwidth: The highest signal frequency that a specific video signal can reach. The higher the video bandwidth, the better the picture quality . A video recorder that can produce a broad video bandwidth generates a detailed, high-quality picture on the screen. Video bandwidths used in studio work vary between 3 and 12 MHz.

Video distribution amplifier (VDA): A special amplifier for strengthening the video signal so that it can be supplied to several video monitors at the same time.

Video equalization corrector (video equalizer): A device that corrects for unequal frequency losses and/or phase errors in the transmission of a video signal.

Video framestore: A device that enables digital storage of one or more images for steady display on a video monitor.

Video gain: The range of light-to-dark values of the image that is proportional to the voltage difference between the black-and-white voltage levels of the video signal. Expressed on the waveform monitor by the voltage level of the whitest whites in the active picture signal. Video gain is related to the contrast of the video image.

Video inline amplifier. A device providing amplification of a video signal.

Video matrix switcher (VMS): A device for switching more than one camera, VCR, video printer, and similar to more than one monitor, VCR, video printer, and similar. Much more complex and more powerful than video switchers.

Video monitor: A device for converting a video signal into an image.

Video printer: A device for converting a video signal to a hard-copy printout. It could be a monochrome (b/w) or color printout. They come in different format sizes. Special paper is needed.

Video signal: An electrical signal containing all the elements of the image produced by a camera or any other source of video information.

Video switcher: A device for switching more than one camera to one or more monitors manually, automatically, or on receipt of an alarm condition.

Video wall: A video wall is a large screen made up of several monitors placed close to one another, so when viewed from a distance, they form a large video screen or wall.

VITS: Video insertion test signals. Specially shaped electronic signals inserted in the invisible lines (in the case of PAL, lines 17, 18, 330, and 331) that determine the quality of reception.

VLF: Very low frequency. Refers to the frequencies in the band between 10 and 30 kHz.

VMD: Video motion detector. A detection device generating an alarm condition in response to a change in the video signal, usually motion, but it can also be due to a change in light. Very practical in CCTV because the VMD analyzes exactly what the camera sees (i.e., there are no blind spots).

VOD: Video on demand. A service that allows users to view whatever program they want whenever they want it with VCR-like control capability such as pause, fast forward, and rewind.

VR: Virtual reality. Computer-generated images and audio that are experienced through high-tech display and sensor systems and whose imagery is under the viewer's control.

VS: Vertical sync.

WAN: Wide area network.

Waveform monitor: Oscilloscope used to display the video waveform.

Wavelet: A particular type of video compression that is especially suitable for CCTV. Offers higher compression ratio with equal or better quality than JPEG.

White balance: An electronic process used in video cameras to retain true colors. It is performed electronically on the basis of a white object in the picture.

White level: This part of the video signal electronically represents the white part of an image. It resides at 0.7 V from the blanking level, whereas the black part is taken as 0 V.

Wow and flutter: Wow refers to low-frequency variations in pitch, whereas flutter refers to high-frequency variations in pitch caused by variations in the tape-to-head speed of a tape machine.

W-VHS: A new wide-VHS standard proposed by JVC, featuring a high-resolution format and an aspect ratio of 16:9.

Y/C: A video format found in Super-VHS video recorders. Luminance is marked with Y and is produced separate from the C, which stands for chrominance. Thus, an S-VHS output Y/C requires two coaxial cables for a perfect output.

Y, R-Y, B-Y: The general set of component video signals used in the PAL system as well as some other encoder and most decoder applications in NTSC systems; Y is the luminance signal, R-Y is the first color difference signal, and B-Y is the second color difference signal.

Y, U, V: Luminance and color difference components for PAL systems; Y, B-Y, R-Y with new names; the derivation from RGB is identical.

Z: In electronics and television this is usually a code for impedance.

Zoom lens: A camera lens that can vary the focal length while keeping the object in focus, giving an impression of coming closer to or going away from an object. It is usually controlled by a keyboard with buttons that are marked zoom-in and zoom-out.

Zoom ratio: A mathematical expression of the two extremes of focal length available on a particular zoom lens.

Index

2.4GHz systems, 184
12 volts DC, 78-82
    AC voltage comparison, 79
    advantages, 80
    camera operation, 79
    common power source, 81
    multiple-circuit power supplies,
      82
    possibilities, 78
    power supplies, 80
    voltage drops, 80
    See also Power
24 volts AC, 82-83
    advantages, 82
    fiber-optic cable and, 83
    multiple outputs, 82
    uses, 83
    voltage drop, 80
    See also Power
110/220 volts AC, 83-84
    camera installation, 84
    drawbacks, 84
    See also Power
900MHz systems, 183-84

Accessibility, 199-207
    camera planning and, 199
    installation, 200

installation-ready dome poles, 203
pole-mount cameras, 200-203
roof mounts, 203-6
service and maintenance, 206-7
See also Outdoor considerations
Advanced planning, 25
Alarm devices, 140-41
    choosing, 224
    configuring, 142
    connecting, 143
    in design stage, 220
Alarm inputs, 145-46
    matrix switchers, 108
    multiplexers, 46-47, 102
    PTZ dome cameras, 68
    quads, 95
    recorders, 46, 139
Alarm-triggered recording, 115-17
Alarm triggers, 134, 139-41
    defined, 139
    devices, 140
    door contacts as, 140, 145
    events, 134
    false, 148
    motion detectors, 141, 146
    for recording speed increase, 45-46
    timer, 146
Analog recording, 111-18
    alarm-triggered, 115-17

changing, to digital, 132-33
continuous, 114-15
event-triggered, 115-17
tape-organizing systems, 118
videotape archiving, 117-18
VTRs, 111-13
See also Recording
Artificial light sources, 196-97
fluorescent, 197
incandescent, 196
mercury vapor, 196-97
neon, 196
Audio recording
covert cameras with, 164-65
in facility, 165
Automatic gain control (AGC), 86-87
Automatic light compensation (ALC),
86-87

Backlight compensation, 87
Bayonet-Neil-Concelman connectors
(BNCs), 173-74
Black-and-white cameras, 58
Broad cameras, 62-63
defined, 62
illustrated, 62
sizes, 63
types of, 62
uses, 62-63
See also Cameras
Bullet cameras, 63-65
connection method, 65
defined, 63
drawback, 64
illustrated, 64
for outdoor applications, 63
uses, 64
See also Cameras

Cabling
choosing, 224-25
coaxial, 167-73
in design stage, 220-21
fiber-optic, 175-78
network video, 180-82
twisted-pair, 178-80
See also Connectivity
Camera locations
day care facilities, 17
hotel/lodging, 12
manufacturing/industrial, 19
office, 14
retail, 9-10
school, 16
Cameras
additional, 28, 30
automatic gain control (AGC), 86-
87
automatic light compensation
(ALC), 86-87
backlight compensation, 87
broad, 62-63
bullet, 63-65
choosing, 57-88, 222-23
color, 14, 57-61
cooling fans, 190
for day care facilities, 17
day/night, 197-98
digital signal processing (DSP),
87-88
dome, 65-69
electronic light compensation
(ELC), 86
features, 86-88
full-size, 69-70
for gaming industry, 20-22
hidden, 157-64
for hotel/lodging, 12-13
layout, 219

for manufacturing/industrial
environment, 17-20
monochrome, 57-61
network, 70-71
office environment, 13-15
pan/tilt zoom, 10, 22, 43
pole-mount, 200-203
power requirements, 78-86
for retail, 6-11
school, 15-16
as silent observer, 2
spare, 33
types of, 61-71
wireless, 183
Camera specifications, 71-78
defined, 71
image sensor, 72
minimum illumination, 75-76
operating humidity, 77
operating temperature, 77
operating voltage, 78
readying, 71
resolution, 73-74
scanning system, 73
S/N ratio, 76-77
video output, 74-75
Camera systems
deferring crime, 4, 5
as deterrent, 4, 5
employee attitudes towards, 4-5
evolution, 1-2
insurance companies and, 4
purpose of, 1-23
as tools, 3
Camera tours, 108, 154·
Card cage, 106
CCTV systems
defined, 1
enhancement, 2
See also Cameras; Camera systems

Changing needs, evaluating, 229-30
Charge-coupled device (CCD) image
sensor, 72
Climate, 187-92
cold environments, 189
harsh environments, 187-88
hot environments, 189-90
humidity extremes, 192
saltwater environments, 190-91
seasonal considerations, 191-92
See also Outdoor considerations
Closed-circuit television video system. *See* CCTV systems
Coaxial cable, 167-73
defined, 167
dielectric, 167
RG/6, 172-73
RG/11, 173
RG/59, 169-72
RG/175, 168-69
shield, 167
See also Connectivity
Coaxial cable connectors, 173-75
BNC, 173-74
crimp/solder, 174-75
twist-on, 174
See also Connectivity
Cold environments, 189
Color cameras, 14, 57-61
choice of, 61
for identification, 60-61
low light levels and, 193
quality misconception, 59
upgrading to, 59
See also Cameras
Connectivity, 167-85
coaxial cable connectors, 173-75
coaxial cabling, 167-73
fiber-optic cabling, 175-78
microwave links, 184

network video cabling, 180-82
twisted-pair cabling, 178-80
wireless video transmission, 182-84
Continuous recording, 114-15, 138
Control equipment, 89-109
    choosing, 223-24
    location, 219
    matrix systems, 104-9
    multiplexers, 96-104
    quads, 93-96
    video switchers, 89-93
Control room layout, 38
Covert cameras. *See* Hidden cameras
Crimp/solder connectors, 174-75

Day care facilities, 17
Day/night cameras, 197-98
Design stage, 219-21
    alarm devices, 220
    camera layout, 219
    control equipment location, 219
    end-user review, 221
    system cabling, 220-21
Digital recording, 118-32
    advantages, 119, 136
    alarm events, 131
    analog conversion to, 132-33
    combination storage systems, 128-29
    digital tape, 126-27
    DV tape, 127-28
    file formats, 120-24
    growth of, 118
    hard-drive, 125-26
    hard drive transfer, 131
    mean time between failure (MTBF), 130
    multiplexing recorders, 129

recorders, 119
signal conversion, 120
single-channel, 129-30
storage types, 124
video as evidence, 132
See also Recording
Digital tape recording, 126-27
    advantages, 127
    archiving, 127
    defined, 126
    recording times, 126
Directional video motion detection, 149-51
    defined, 150
    programming options, 151
    uses, 151
    See also Video motion detection
Dome cameras, 65-69
    fixed, 65-67
    PTZ, 67-69
    See also Cameras
Door contacts, 140, 145
Dual-page quads, 96
Duplex multiplexers, 99-103
    defined, 100
    recording/playback, 100
    with spot monitor, 100
    use of, 100-101
    See also Multiplexers
DV tape recording, 127-28
Dwell time, 90-91

Electronic light compensation (ELC), 86
Emergency Call station, 214
Enhanced recording, 133-34, 137-55
    alarm inputs, 145-46
    alarm triggers, 134, 139-41
    frame frequency, 143-45

with PTZ cameras, 151-55
  recording speed, 141-43
  reviewing techniques, 137-39
  video motion detection, 146-51
  See also Recording
Equipment selection stage, 221-25
  alarm device selection, 224
  cabling/equipment selection, 224-25
  camera selection, 222-23
  control equipment selection, 223-24
  end-user review, 225
Event-triggered recording, 115-17
Event-triggered review, 47-48
  applications, 47
  random review with, 50
  See also Review
Exclusive recording, 143-44
Expandability, 28-31
  components, 28
  matrix switcher, 28-30
  multiplexer, 30
  See also Requirements

Fiber-optic cabling, 175-78
  configurations, 175
  defined, 175
  in design stage, 220-21
  drawbacks, 175-76
  module, 176
  PTZ camera and, 177-78
  transmitters and receivers, 177-78
  trunks, 177
  See also Connectivity
File formats, 120-24
  H-263, 123-24
  JPEG, 121
  MPEG, 121-23

wavelet, 123
  See also Digital recording
Fixed dome cameras, 65-67
  defined, 65-66
  illustrated, 66
  large, 67
  outdoor, 67
  small, 66
  See also Cameras; Dome cameras
Fluorescent light, 197
Frame frequency, 143-45
Frame rate
  defined, 11
  increased, 45
  VCR, 11
Full-size cameras, 69-70
  camera mounts, 70
  illustrated, 69
  options, 70
  outdoor, 70
  zoom lenses, 70
  See also Cameras

Gaming industry
  cameras for, 20-22
  camera types, 22
  design considerations, 21-22
  key factors, 21
  number of cameras, 22
  video surveillance, 21
Grab-and-run, 9

H-263 compression, 123-24
  defined, 123
  MPEG vs., 124
  video streaming, 124
  See also File formats
Halogen infrared illuminators, 195

Hard-drive recording, 125-26
Harsh environments, 187-88
Hidden cameras, 157-64
   with audio, 164-65
   best application of, 159
   enhancing, 159-60
   illustrated, 158
   liability issue, 159
   in motion detector, 160, 161
   overuse of, 157
   private offices and, 160-64
   smoke detector, 162
   using, 157-60
   See also Cameras
Hotel/lodging security
   camera locations, 12
   cameras, 12-13
   design considerations, 12-13
   recording, 12-13
   tape storage, 13
Hot environments, 189-90
   cooling fans and, 190
   housing protection and, 190
   See also Outdoor considerations
Humidity, 192

Identification, color cameras for, 60-61
Image sensors, 72
Incandescent light, 196
Independent poles, 201-2
Infrared illuminators, 195
Installation. *See* System installation
Installation-ready dome poles, 203
Interleave recording, 144-45

JPEG, 121

LED illuminators, 195-96
Lighting, 192-99
   artificial sources, 196-97
   day/night cameras and, 197-98
   direct sunlight and, 198-99
   levels, 193
   levels, increasing, 195
   levels, measuring, 194-95
   low levels, 193-96
   See also Outdoor considerations
Light meters, 193-94
Lipstick cameras. *See* Bullet cameras
Looping inputs, 91
Low light levels, 193-96
   color cameras and, 193
   increasing, 195
   See also Lighting

Manufacturing/industrial environ-
    ment
   camera locations, 19
   cameras for, 17-20
   control equipment location, 19-20
   design considerations, 18-20
   environmental conditions, 18
   for monitoring process control
    equipment, 20
   recording, 19
   safety approach, 18
Matrix switchers, 28-30, 104-9
   alarm inputs, 108
   as backup, 32
   camera tours, 108
   card cage, 106
   control, 106-7
   defined, 28, 105
   environments, 28-29
   expandability, 105-6
   functionality, 107

illustrated, 29, 30, 106
monitor outputs, 29-30
remote keyboard control, 106-7
size, 105-6
subject tracking, 108
video inputs, 29
See also Control equipment
Mean time between failure (MTBF),
130
Mercury vapor light, 196-97
Microwave links, 184
Minimum illumination, 75-76
defined, 75
f-number, 75-76
See also Camera specifications
Monitoring employees, 213
Monitors
arrangement, 36, 37
monochrome, 60
rotating, 32-33
size, 36-37
in viewing, 35-36
Monochrome cameras, 57-61
defined, 58
picture clarity, 60
price, 59
use of, 59
See also Cameras
Motion detectors, 141, 146
hidden camera in, 160, 161
multiple, 146
video, 146-51
MPEG, 121-23
defined, 121
drawbacks, 122
H-263 standard vs., 124
MPEG-1, 122
MPEG-2, 122-23
MPEG-4, 123
types of, 122-24

See also File formats
Multiplexers, 28, 96-104
alarm inputs, 46-47, 102
choosing, 101
color, 59, 101-2
controls, 103-4
defined, 30, 96-97
duplex, 99-103
example use of, 98
illustrated, 97
monochrome, 101-2
multiscreen views from, 99
options, 101-2
reasons for adding, 30
recorder output, 97
relay outputs, 102-3
remote, keyboard, 104
simplex, 99-103
video motion detection, 103

National Labor Relations Board
(NLRB), 160
Needs analysis, 217-18
defined, 217
parking garage, 217-18
Network cameras, 70-71
defined, 70-71
development, 71
disadvantage, 71
in traditional camera system, 181
See also Cameras
Network video cabling, 180-82
NTSC, 74

Office environment
camera locations, 14
cameras, 13-15
camera types, 14

design considerations, 14-15
recording, 15
theft, 13
violence, 13
Operating humidity, 77
Operating temperature, 77
Operating voltage, 78
Outdoor considerations, 187-207
accessibility, 199-207
climate, 187-92
equipment selection and, 188
lighting, 192-99

PAL, 74
Pan/tilt zoom cameras
in casinos, 22
controlling, 178
domes, 67-69
enhanced recording with, 151-55
fiber-optic cabling and, 177-78
outdoor, 152
presets, 151-52
presets, utilizing, 153-55
problems, 151
in retail locations, 10
use of, 43
See also Cameras
Parapet mounts, 206
defined, 206
illustrated, 204
uses, 206
See also Roof mounts
Periodic review/analysis, 229-30
Picture quality, 60
Picture resolution, 73-74
Planning stage, 210-18
monitoring employees, 213
needs analysis, 217-18
questions, 215

security measures, 213-15
site survey, 215
survey from drawings, 215-16
survey from drawings and site
visit, 216-17
survey on site, 216
user goals, 212-13
user survey, 211-12
Playback, 47-55
Pole-mount cameras, 200-203
existing poles, 200-201
independent poles, 201-2
pan/tilt units, replacing, 202-3
See also Accessibility; Cameras
Power
12 volts DC, 78-82
24 volts AC, 80, 82-83
110/220 volts AC, 83-84
AC and DC voltage comparison,
79
centrally-controlled, 85
requirements determination, 78-86
summary, 85-86
Presets, 151-52
camera park, 152
triggers, 153
utilizing, 153-55
See also Pan/tilt zoom cameras
Private office hidden cameras, 160-64
California case, 162-63
criticism, 161
NLRB ruling, 160
smoke detector, 162
See also Hidden cameras
PTZ dome cameras, 67-69
alarm inputs, 68
defined, 67
features, 67-68
illustrated, 68
outdoor, 69

preposition settings, 68-69
receiver, 67
See also Dome cameras

Quads, 93-96
    advantages, 93
    alarm inputs, 95
    for color/monochrome systems,
       95
    defined, 93
    display options, 96
    dual-page, 96
    full-duplex, 94
    illustrated, 94
    See also Control equipment

Random review, 50-51
Real-motion recorders, 113-14
    defined, 113
    on-screen recording schedule, 114
Recorders, 97
    alarm input, 46, 139
    alarm-triggered, 134
    digital, 119
    digital multiplexing, 129
    digital single-channel, 129-30
    DV, 127
    illustrated, 112
    locked, 54
    manufacturing/industrial envi-
       ronment, 19-20
    real-motion, 113-14
    relay outputs, 139
    retail security, 11
    time-lapse (VTRs), 111-13
Recording, 44-47, 111-36
    after-hours, 11
    alarm-triggered, 115-17

amount, 45
analog systems, 111-18
audio, 164-65
buffer, 134
capabilities control, 54
continuous, 114-15, 138
day care facilities, 17
digital systems, 118-32
enhanced capabilities, 133-34, 137-
    55
event-triggered, 115-17
exclusive, 143-44
hard-drive, 125-26
hotel/lodging, 12-13
importance, 44
interleave, 144-45
manufacturing/industrial, 19
office environment, 15
rate, changing, 142-43
real-time, 11
retail, 11
schools, 16
speed, 11
speed, changing, 141-43
storage types, 124
time, void, 32
See also System elements
Redundancy, 31-34
    cameras, 33
    connections, 32
    monitors, 32-33
    See also Requirements
Refresh rate, 98
Relay outputs, 102-3
Requirements, 25-34
    advanced planning, 25
    expandability, 28-31
    redundancy, 31-34
    threat assessment, 25-28
Resolution, 73-74

Retail environments
    camera locations, 9-10·
    cameras for, 6-11
    checkout lanes, 7
    design considerations, 9-11
    pan/tilt zoom cameras, 10
    recorders, 11
    scams, 8
    warehouse areas, 8
Return theft, 7-8
Review, 47-55
    event-triggered, 47-48
    locations, 52-53
    method choice, 51-52
    methods, 47-51
    personnel, 54-55
    random, 50-51
    times, 53-54
    total, 48-49
Reviewing techniques, 137-39
RG/6 coaxial cable, 172-73
RG/11 coaxial cable, 173
RG/59 coaxial cable, 169-72
    BNC and RG/175 cable comparison, 171
    distance, 169
    illustrated, 169
    riser-rated coax, 172
    solid-center conductor, 171-72
    stranded-center conductor, 171
    types of, 171-72
    See also Coaxial cable
RG/175 coaxial cable, 168-69
Risk analysis. *See* Threat assessment
Robberies
    grab-and-run, 9
    returns, 7-8
    smash-and-grab, 8, 9
Roof mounts, 203-6
    parapet mount, 206

    pole mount, 104
    types of, 204
    wall mount, 204-6
    See also Accessibility

Saltwater environments, 190-91
    camera life spans and, 191
    protection, 190
Scanning system, 73
School security
    camera choice, 16
    camera locations, 16
    cameras for, 15-16
    design considerations, 15-16
    recording, 16
Security
    hotel/lodging, 12-13
    manager, 54-55
    measures, 213-15
    office, 13-15
    retail, 6-11
    school, 15-16
Service and maintenance access, 206-7
Simplex multiplexers, 99-103
    choosing, 101
    defined, 99-100
    previously recorded video playback, 100
    See also Multiplexers
Site survey, 215
Smash-and-grab, 8
S/N ratio, 76-77
Sunlight, 198-99
Survey
    from drawings, 215-16
    from drawings and site visit, 216-17
    on site, 216
System elements, 35-55

playback and review, 47-55
recording, 44-47
viewing, 35-44
System expansion, 230
System installation, 225-28
    access, 200
    documentation, 228
    managing, 226
    planning, 226
    reviewing/accepting, 226-28
System use, 228-29
    technical training, 228-29
    user training, 229

Tape-organizing systems, 118
Technical training, 228-29
Threat assessment, 25-28
    care in, 27
    defined, 26
    steps, 26
    system design and, 27-28
Total review, 48-49
    defined, 48
    frequency, 49
    planning for, 48-49
    See also Review
Transmitters/receivers
    2.4GHz system, 184
    900MHz system, 183
    fiber-optic, 177-78
    twisted-pair (active), 179-80
    twisted-pair (passive), 179
    wireless, 182-83
Twisted-pair cabling, 178-80
    active transmitters and receivers,
        179-80
    alternate configurations, 180
    passive transmitters and receivers,
        179

transmission over, 221
    See also Connectivity
Twist-on connectors, 174

Users
    goals, 212-13
    survey, 211-12
    training, 229

Video motion detection, 103, 146-51
    advantages, 147
    area setup, 149
    defined, 147
    detection zones, 149
    directional, 149-51
    false triggers, 148
    sensitivity levels, 149
Video multiplexing, 2
Video output, 74-75
    BNC, 74-75
    cable selection and, 75
    defined, 74
    impedance matching, 74
    NTSC/PAL, 74
    See also Camera specifications
Video switchers, 89-93
    defined, 89
    illustrated, 92
    looping inputs, 91
    options, 91-93
    second monitor output, 92
    See also Control equipment
Videotapes, archiving, 117-18
Video time-lapse recorders (VTRs),
    111-13
    in 24-hour mode, 112
    in 72-hour mode, 112-13
    defined, 111

modes, 111
    on-screen recording schedule, 114
Viewing, 35-44
    control room, 38
    monitors, 35-37
    previously recorded tapes, 46
    proper, 42
    setup illustration, 36
    in smaller facilities, 37
    See also System elements
Viewing area, 39-44
    actual size, 41
    determining, 40-41
    focused, 42
    illustrated, 40
    limits, 40-41
    viewing subject size, 41-42
Viewing subject, 39
    changing, 43
    determination, 40
    size, 41-42

Wall mounts, 104-6
    defined, 204-5
    illustrated, 205
    installation, maintenance, repair,
        206
    See also Roof mounts
Warranties, 34
Wavelet technology, 123
Wireless video transmission, 182-84
    2.4GHz systems, 184
    900MHz systems, 183-84
    advantages/disadvantages, 182
    long-range transmitters/receivers,
        183
    See also Connectivity

Zoom
    for identity capture, 43
    mistake, 42

About the Author

Alan R. Matchett, C.P.P., is a Security Design Engineer for Duke, Cogema, Stone, and Webster, based in Charlotte, North Carolina, performing system design, design review, and specification preparation and writing for nuclear facilities. With over 18 years of experience, Mr. Matchett has worked in a wide variety of environments and has established himself as a leading authority in CCTV and access control. As a Security Engineer for Dyncorp, based in Reston, Virginia, he has performed security work in nearly 80 countries throughout the world. Mr. Matchett has also performed design, analysis, installation, repairs, and project management for residential, industrial, office environment, retail, military, foreign and U.S. government, educational, healthcare, telecommunication, hotel and lodging, and food service customers.

Mr. Matchett previously worked as a Security Consultant and Project Manager for Dyncorp, primarily working with the U.S. Department of State. Before returning to Dyncorp in 2000, he founded Eye See U! Security Enterprises and Eyeseeu.com as a security consulting and training firm based in Alexandria, Virginia. Mr. Matchett has a B.A. in creative writing and a B.S. in electronic engineering technology. He has been a Certified Protection Professional since 1993 and is frequently called upon for speaking engagements internationally. He has written numerous articles for security publications, newspapers, and magazines.

Printed and bound by CPI Group (UK) Ltd, Croydon, CR0 4YY

03/10/2024

01040335-0014